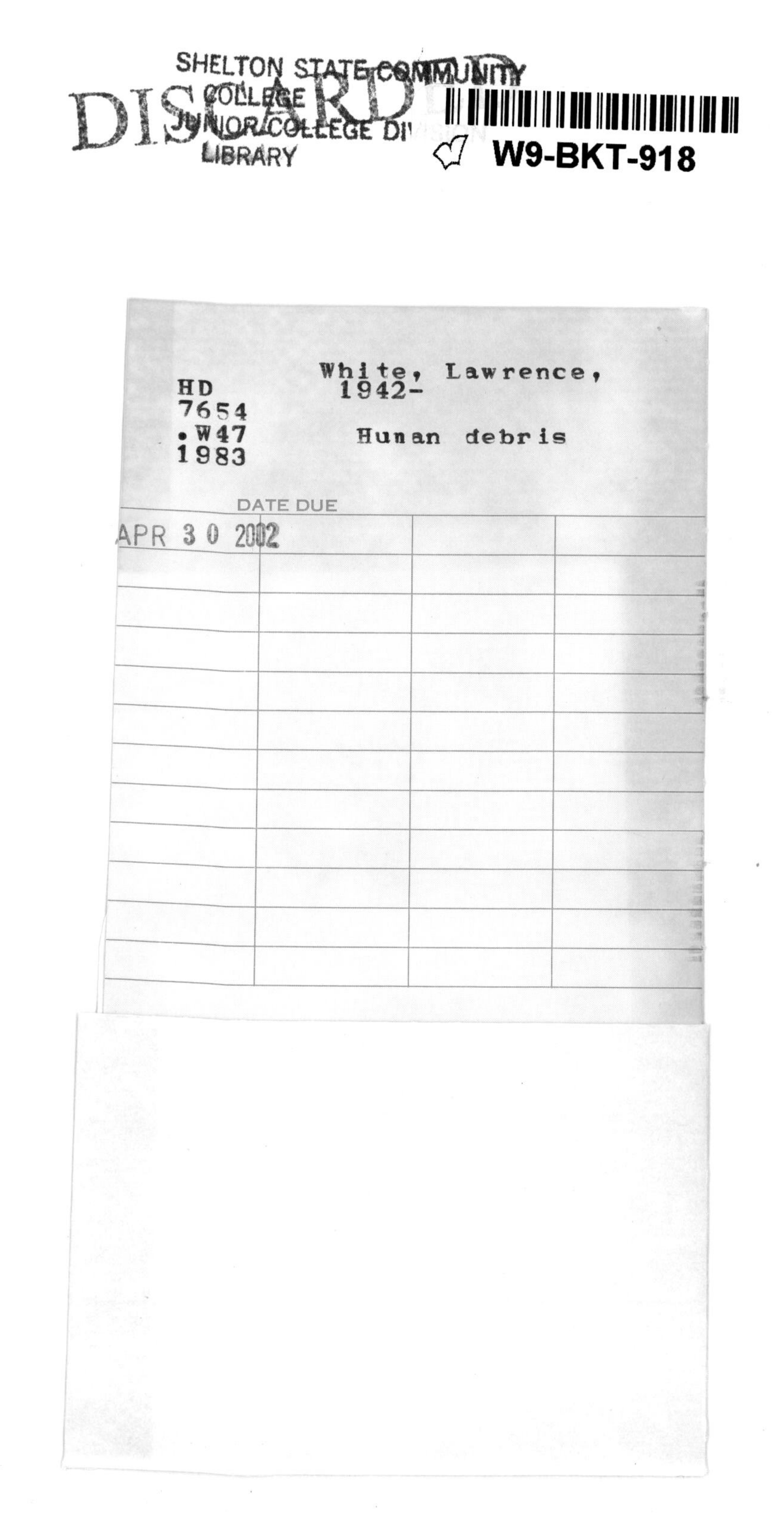

White, Lawrence, 1942-

Hunan debris

HUMAN DEBRIS

HUMAN DEBRIS

The Injured Worker in America

Lawrence White

SEAVIEW/PUTNAM
NEW YORK

for Jane

 Published simultaneously in Canada by General Publishing Co., Limited, Toronto.

The text of this book is set in 11 point Electra

Library of Congress Cataloging in Publication Data

White, Lawrence, date.
Human debris.

1. Industrial hygiene—United States. 2. Industrial accidents—United States. 3. Workers' compensation—United States. I. Title.
HD7654.W47 363.1'1'0973 81-84540
ISBN 0-399-31013-4

PRINTED IN THE UNITED STATES OF AMERICA

Contents

Preface and Acknowledgments

This book is first of all an exposé of a decaying corner of our society. The truth of the social problems dealt with in *Human Debris* is nothing less than unnecessary human suffering on a vast scale. The stories of injured workers in Chapters 1, 2, 3, and 6 show the effect on the lives of real people of our country's neglect of the problems of industrial injury, disease, and workers' compensation.

It would be enough, I believe, to read these chapters in order to gain a basic understanding of the problems of on-the-job injury and compensation. For those who wish to probe further there is a great deal more. Chapters 4, 5, and 7 taken together explain the essential features of the workers' compensation system. Chapter 8 traces the history of the Occupational Safety and Health Act and its implementation by the Reagan administration. Chapter 9 deals with the anachronistic but oddly fashionable issue of state versus federal authority. The last chapter contains some concrete recommendations for positive change.

Specialists in occupational safety and health and workers' compensation will find much that is familiar to them. They may also learn something new. For too long these two topics have been treated separately and experts in one field have lost touch with the other. I hope that activists in the occupational safety and health field take this opportunity to learn about workers' compensation. Industrial accidents and disease and compensation for them are interrelated phenomena. If health and safety conditions for American workers are to improve, reform in one area must move in tandem with reform in the other.

In many places in this book I have advocated the injured worker's right to sue his employer. I believe that the stripping away of this right by early workers' compensation laws was an egregious deprivation of a basic civil right. Nevertheless, I recognize the inadequacies of the civil litigation system in our country. Civil lawsuits are time-consuming and the results are uncertain. They cannot and should not substitute for a rational compensation system. But that system should not immunize the employer from the consequences of negligence or misconduct in injuring workers. While the civil courts are a deterrent—perhaps the most effective deterrent—to the careless employer, they cannot provide the swift and certain compensation that injured workers need.

No book covering such a large subject can be done without the help of many persons, all of whom I'd like to thank. First of all, I'm grateful to Kim Chernin, without whose encouragement and advice this book would not have been written. I also want to thank Dan Berman, who generously shared his extensive knowledge of occupational health and safety issues and guided me through the network of activists. Mary Jones, formerly of the Cook County Hospital Occupational Health Clinic, was a valuable source of referrals to injured workers. Patrick McGuire of the Chicago Committee on Occupational Safety and Health (CCOSH) provided me with important information and encouragement. Ken Lewis and David Hollenbeck were two dedicated lawyers who understood the need for this book. John Lawrence, staff director, and Michael Goldberg, chief counsel, of the Labor Standards Subcommittee of the House of Representatives Labor Committee, were always ready to help. Scott Lilly gave me valuable insight into the mentality of the current administration in Washington. Allan Tebb helped me understand the arcane world of workers' compensation insurance, as did Robert E. Meyer. Other people without whose help this book would have been less than it is are: Dr. Eula Bingham, Paul Cossaboon,

Dr. Molly Coye, Jack Benjamin, Terri Gerritz, Mark Greenberg, Susan Griffin, Joe Hernandez, Robert Hunter, Anne Isaak, Dr. Gary Kelsberg, Prof. Edward F. Mooney, Anne Pilsbury, Peggy Semenario, Jack Sheehan, Warren Smith, Joyce Tom, Joe Velasquez, Len Welsh, and Michael Wilk. My thanks also to the many other generous persons I haven't mentioned who have helped me with their thoughts and their time.

I regret that it was not possible for me to chronicle the activities of the individuals and groups who have taken it upon themselves to champion the interests of the health and safety of our nation's workers. Their efforts have never been more important than they are at this moment.

Berkeley, 1982

HUMAN DEBRIS

1

Broken Backs and Dried-Up Lungs

Industrial injury, occupational disease, and workers' compensation have not been newsworthy in the conventional sense since the early 1900s. Except for the cataclysmic event—the collapse of a building under construction or a mine explosion—little media attention focuses on the American workplace and its dangers. It is thought that most on-the-job injuries are too common, too much a part of ordinary life, to be considered interesting to the general public, while occupational disease is esoteric and undramatic. When a major accident occurs, the story that is in the headlines one day is forgotten the next. But what really happens to the survivors of those accidents, after the fickle light of media attention has moved on to someone else?

Consider the case of a construction worker I'll call Bill O'Leary. Like all the workers in this book, his name has been changed, but the details of what happened to him have not been altered. Bill was on the front page of the Boston papers one day in 1970. He had been working on the ground beneath the huge Harbor Towers complex when a plank, 4″ by 4″ and twelve feet long, fell from twenty-eight stories above him and broke on his head. Bill was especially newsworthy because he survived. He was six feet two and a half inches tall and weighed 240 pounds; he had played semipro football for seven years before the injury. A year after the injury he was six feet one-half inch tall and weighed 115 pounds. The doctors explained that his body had contracted during the year that he'd been in a coma.

Bill was in the news again once or twice—human interest stories; Bill coming out of a coma—no one thought it was possible; Bill coming home from the hospital to his wife and five small children.

Not much was reported of the background of the accident; that objects had been falling from the top of this construction project so often that the secretary-treasurer of the building trades council himself had asked the company to construct safety nets five or six times to no avail. That on the day of the accident, a Friday, no one should have been working directly under a work crew on top of the tall building but that the contractor was anxious to finish the job and didn't want to pay double time for Saturday work.

Twelve years after the accident, Bill is no longer "newsworthy." His head is literally patched together from other parts of his body. The hair growing over the front of his head is scanty and oddly coarse. It used to grow on his chest—some of the skin from there was transplanted onto his head. A muscle from under his arm was also transplanted onto his head. There are stitch marks on Bill's skull where the surgeons have done their best to match patches of artificial materials covered with transplanted tissue with fragments of Bill's skull. Extremely lucky to be able to walk and talk, which he does with some effort, Bill is unable to do any kind of work. When he was injured he was only thirty years old; he is now forty-two.

Because he was injured on the job, Bill was covered by workers' compensation. There are several implications to this fact. First, no matter how negligent his employer was, Bill could not sue in civil court. Workers' compensation is the "exclusive remedy," insulating employers from their own employees' lawsuits for on-the-job injuries. Bill must accept the amount of benefits set by state law. Bill and his family of seven now receive, as they have received since the accident, $106 per week in Workers' Compensation, the maximum allowed under Massachusetts law at the time of Bill's injury.

The hazards of working in America are far worse than most people realize. Bill O'Leary was lucky to be alive. He didn't become part of the number 14,000, which is the National Safety Council's estimate of workers killed on the job each year. Instead he became part of the 2.2 million disabled as a result of industrial accidents *each year.* Even these figures are probably too low. Nicholas Ashford, one of the leading commentators in the field of occupational safety and health, says that the 2.2 million is low by a factor of five. *

Even more ominous is the problem of occupational disease. The U.S. Department of Labor in a 1980 report found that *over 100,000 Americans per year die of an occupational disease!* Of these, over 50,000 will die each year for the next few decades as a result of exposure to asbestos. A study done by Dr. Irving Selikoff, the country's leading occupational-health researcher, found that of insulation workers and shipyard workers with substantial exposure to asbestos, 20% to 25% died of lung cancer, 10% to 18% died of asbestosis, and 10% died of gastrointestinal cancer.

A controversial Health, Education and Welfare Department study published in 1978 concluded that past exposure to occupational carcinogens may account for as much as 23% to 38% of all cancer deaths over the next several decades. And the specter of occupationally induced cancer is looming ever larger. The virtually unlimited introduction of new chemicals in the workplace endangers large segments of the working population.

The worker who develops cancer from exposure to asbestos or some exotic chemical and the worker who is hit on the head by an object falling from above have a major institution in common. Workers' compensation is their sole recourse and the only economic deterrent to employers intent on getting a job done regardless of consequences to workers.

Joseph Rawlings' case was less dramatic than Bill O'Leary's. His back injury represents the single most common type of work-related disability. He had been a metal fabricator, something akin to a carpenter with metal, when he injured his back. A 300-pound hunk of metal pipe he had been manhandling almost fell on him. To save his foot he yanked the pipe and it fell to the side. Rawlings' back was

* Nicholas Ashford. *Crisis in the Workplace: Occupational Disease and Injury.* First edition, 2nd printing. Cambridge, MA: MIT Press, 1976, p. 3.

never the same. As soon as it became apparent that he couldn't come back to work for an indefinite time, his boss fired him. Six months later Rawlings felt that he could work again and he got a job in another metal-fabricating shop. Almost a year later he was helping someone move a very heavy drum filled with a solvent when he felt a snapping sensation in his back. He finished his shift, but felt very stiff. That night he woke up in horrible pain and his long journey through misery began.

Joe Rawlings' back was broken. The doctors, orthopedists, had scientific names for what happened to him, and they disagreed about exactly what to call his condition, but he was in fact broken. He would never again have the strength or stamina to lift or carry, to do the tasks he had to do to be a metal fabricator or do any manual work. Indeed, he could no longer pick up his two young children or mow his lawn. His sex life was gone, if not permanently, at least for several years. For people with bad backs, sex is more pain than pleasure.

Rawlings went to a lawyer because his workers' compensation settlement had been delayed and he didn't understand why. He eventually found out that two workers' compensation insurance carriers were responsible for his injuries, one for each of his employers, and they were fighting each other about who would pay what. In the meantime, Joe had to wait. When he was hurt, he didn't even know that workers' compensation was private insurance. Like most people, he thought it was run by the state, like unemployment insurance.

Joe had been getting $154 per week after his back injury disabled him. This was about 40% of his take-home pay. The older of his two children was three. Eve Rawlings had never worked and, like her husband, had no education after high school. They couldn't make ends meet.

One day when Joe was out of the house, Eve called his lawyer. Her voice was urgent as she told him of their financial predicament. She didn't want Joe to know that she had called. He was very proud and would not ask for favors or special treatment. The problem was that they were four months behind in their house payments and the bank was threatening to foreclose. Was there any way they could get some money in advance from the insurance company? Joe was entitled to some amount of money based on the degree of disability that he had suffered. He'd been waiting for that sum for several years now. If only the Rawlings could get some part of it they could save their house.

Being young and inexperienced, their lawyer felt optimistic about being able to help. He first called the bank and convinced them to hold off for another two weeks. Then he called the two workers' compensation insurance carriers and discussed the case with several claims supervisors. They all agreed that Rawlings was entitled to far more than was needed to pay the past due house payments. But as for any advances, no, they had already given him two advances and they wouldn't give any more. Clearly, it was not Rawlings' fault that the case had dragged on so long; the dispute was between two insurance carriers, but, sorry, there was nothing more they could do.

The date for the hearing at the workers' compensation board, a state agency that settles disputes between injured workers and workers' compensation carriers, and the only power that could make the carriers pay, was set for two months hence. It could not be moved forward. Repeated phone calls and letters to the insurance companies didn't help. The Rawlings lost the house they had worked so hard to buy.

There is nothing unusual about Joe Rawlings' experience. There are millions of Americans injured on the job whose lives have been destroyed. Before they were injured, they had thought that workers' compensation would take care of them for their work-related injuries—if they thought about the subject at all. It was only afterward thay they learned how little was available to help them and how grudgingly it was given. The best lawyers tell their clients not to expect any real help from workers' compensation. As terrible as it is to face rebuilding a life with a broken body and no outside help, it is even worse to be waiting for help from a source that will inevitably prove useless.

The vast majority of workers are covered by state workers' compensation programs. Of the $11.9 billion in workers' compensation benefits paid in 1979, $9.5 billion was paid in the fifty states and the District of Columbia and $2.4 billion was paid under federal laws. The Social Security Administration estimates that 78.6 million

American workers were covered by workers' compensation insurance in an average month in 1979. There are several federally governed programs of workers' compensation, but they are directed at special employee groups representing a small percentage of the nation's workforce. For example, the Longshoremen's and Harbor Workers' Compensation Act covers some people who work on the waterfront. Federal employees are covered by their own compensation act, while railroad workers engaged in interstate commerce are not covered by workers' compensation at all but by special employers' liability act. Formerly the elite of American workers, these railroad employees have retained the right to sue their employers despite many attempts by the railroads to try to force workers' compensation on them.

Coal miners suffering from "black lung," the disease characteristic of their occupation, are compensated by a special federal program instituted for their benefit and administered differently from other workers' compensation programs. While most workers' compensation is financed by insurance premiums paid by employers to private insurance companies, the black lung program is financed by a tax on coal paid by mine owners to the federal government.

If railroad employees are in a sense above workers' compensation, farmworkers are below it. Mostly uneducated and without access to lawyers, the nation's farmworkers cannot as a practical matter sue their employers for any injuries they receive on the job as a result of the employer's negligence. At the same time, many state workers' compensation laws specifically exclude farmworkers from coverage. The farmworker whose legs are broken when his tractor overturns because the farmer's road is full of holes has nothing to turn to—the charity hospital and possibly welfare, if he's lucky. This is especially alarming in light of the fact that farm work is the third-most hazardous occupation—less dangerous than only mining and construction work.

The vast majority of work-related injuries are mundane. It is perhaps more exciting to focus on the truly horrible and the exotic; once in a while a newspaper will pick up such a story and the public will be treated to a photograph of a man who had been overcome by toxic fumes at work and been brain damaged. There will be a picture of his wife sitting next to his bed, where he is lying, blank faced, a human vegetable. There are many such stories, enough for thousands of newspaper stories. But there is a danger of being emotionally overwhelmed by them and missing the point. It is the less dramatic, the routine injuries, that happen again and again that demonstrate more clearly that industrial injuries and occupational disease are not just accidents. They are a built-in feature of the system of production in many workplaces.

Take the case of Martha Cook. She worked for Safeway, one of the nation's largest supermarket chains, for twenty years. Mrs. Cook is now fifty-one and no longer works there. Why? Because she is suffering from a condition called "carpal tunnel syndrome," an affliction that is common among supermarket workers, which makes it very painful to press on anything with her right hand. Cook was a checker, pounding on a cash register all day long for fifteen of her twenty years at Safeway. Her condition was caused by years of repetitive strain on her hands and arms.

Her doctor wanted her to return to light duty on a trial basis, slowly increasing the difficulty of work to see how much she could tolerate. Safeway, like many other major employers, has an all-or-nothing policy—come back to work able to do every job available or don't come back at all. So despite her experience, her obvious intelligence, her dedication to hard work (she has reared nine children and kept an immaculate home while working full time for Safeway), despite her three operations to correct the problem, Cook has been fired from Safeway.

Martha Cook earned about $350 per week as a "journeyman food clerk" before she had to stop working. Now she is attempting to sell cosmetic products and is lucky to earn $100 per week. She does not have a high school diploma and has no work experience except Safeway. Mrs. Cook thinks, perhaps rightly, that at fifty-one, with a disability and no formal education to speak of, she will never be able to get a job paying anything like what she was used to. Her attitude is a mixture of resignation and anger—she is not the type of person to be bitter; her children are grown and she has a husband who earns good money. What hurts her most is that she can no longer earn her own way, something which she insisted on doing most of her life.

Workers' compensation offers very little to Martha Cook. She received $154 per week while she was being treated for her hand condition, and the insurance company paid for the operations (although her group medical plan would have paid if the workers' compensation carrier hadn't). Martha received a few thousand dollars in final settlement of her case, since her residual disability was not extensive. The fact that she could not do her job because of this small degree of disability, the fact of her being deprived of her only skill and her job of twenty years, could not be taken into account under the workers' compensation laws.

Carpal tunnel syndrome is becoming epidemic in this country. The National Institute for Occupational Safety and Health (NIOSH) is engaged in a long-term study of the problem. But the type of unrelieved repetitive work that causes the condition continues to be imposed on workers as if no one knew its results. Perhaps in future years, under a different administration, the Occupational Safety and Health Administration (OSHA), the agency in charge of on-the-job safety and health, will attempt to regulate the causes of carpal tunnel syndrome. Until then, employers are free to require their employees to do the jobs that will result in this disabling condition. Indeed, the American worker may be subjected to much worse on pain of losing his job.

Consider the following, a story reported in the San Francisco *Chronicle*, about a Supreme Court case:

"The case began with the refusal of two maintenance workers at a Whirlpool Corp. appliance plant in Marion, Ohio, to step onto a wire mesh screen 20 feet above the factory floor.

"In the months before the two men refused their assignment, several workers had fallen through the screen, one to his death. The two employees had complained to the plant safety director about the screen and had also called the regional office of the Occupational Safety and Health Administration."

The Supreme Court decided that, given those facts, the workers had a right to refuse the job assignment and could not be fired for doing so, although the employer did not have to pay a worker who refused this kind of dangerous task. Furthermore, the worker who refused the job might later be fired if a court found his refusal "unreasonable." This court decision, begrudgingly giving American workers the right to preserve their lives on the job under certain very limited circumstances, was decided not in 1890, but February 26, 1980. It reflects a health and safety status of the American worker which is oddly reminiscent of the dark ages of industrial capitalism.

It is time that the issue of death and injury and disease in the workplace became a subject of major public debate. Not since the early part of the century has this happened in the United States, despite the passage of the Occupational Safety and Health Act in 1970. The public has not really understood the dimensions of the problems and the need for reform.

We should not be too smug about the progress we have made in eighty years. The American worker may be spared some of the cruder hazards of earlier times, but is now subject to the more subtle but just as lethal risks of our high-technology society. A certain percentage, still unknown, of those who work with atomic energy, pesticides, herbicides, asbestos, and a number of other highly toxic substances will die horrible deaths of cancer. Many will never know that it was exposure at work that caused their death, because the time it takes for the cancer to develop may be twenty or thirty years. And there are companies thriving right now that show such a reckless disregard for the lives of their workers that they resemble all too closely the unscrupulous cartels of an earlier time.

The great difference between the beginning of the century and its end is that now it takes more knowledge and imagination to see the worst dangers that confront workers. These dangers are often silent, colorless, odorless, and invisible; dangers like atomic radiation, which the human mind has created and from which only the human mind properly used can preserve us.

Most newly injured workers, like the general public, think that the problems of workplace safety and proper compensation for injury are being "taken care of." They learn through their own bitter experience that nothing is being "taken care of."

The situation at Whirlpool is a metaphor. American workers are asked every day to step out onto a fragile net through which they may fall, now or twenty years later, to their death.

While injured workers are merely a waste product to the industries that have used them up, they are still useful to other sectors of the economy. The workers' compensation insurance industry, a multibillion-dollar enterprise, thrives on them. The more workers are injured the more insurance is needed and the more insurers can charge for the insurance.

Workers' compensation generates a huge amount of litigation, and that benefits lawyers. Since the litigation usually involves issues such as nature and extent of disability or work-relatedness of a disease, medical opinions are needed in bulk. Doctors have a piece of the action. Industrial clinics catering to employers wishing to control their employees' treatment and perhaps have an input into the diagnosis are a new phenomenon mushrooming around the country, giving work to medical entrepreneurs.

Litigation has become such a major aspect of workers' compensation in California, for example, that more money is paid to lawyers for litigating claims than is paid out for medical care for injured workers. Lawyers, doctors, insurance companies—everyone benefits from the present workers' compensation system—everyone except injured workers.

That American workers and their families no longer face death from starvation and the elements as a result of job-caused disability, as they once did, has nothing to do with worker's compensation. It is the federal government through the Social Security disability program and Aid to Families with Dependent Children (AFDC) that keeps the wolf from the door long after workers' compensation has run out.

The "general public," the people who are not themselves exposed to dangers in their own workplaces, have a very real stake in reducing workplace health and safety hazards. The destruction of the lives of millions of our fellow citizens, of their ability to work, to take care of themselves and their families, costs us all dearly. For it is our tax dollars that support most of the industrially disabled, and not the industries that caused the disabilities. There is also a moral cost to our nation that cannot be overlooked. If the people upon whom we all depend for our basic goods and services can be so easily *used up* and made economic outcasts, how can we expect loyalty to a job, to an employer, to a society that allows this to happen? There is a great reservoir of anger and bitterness growing among the millions who have been hurt on the job and thrown on society's ash heap. We ignore it at our peril.

The attitude toward its workers of much of industry is a mirror image of its attitude toward the public. From consumer safety to environmental protection, every advance, every sign of progress, has required a pitched political battle against intransigent vested interests. Workers are in the front line of those exposed to polluted air, toxic substances, radiation. If they are protected, industry would have no argument to support lax health and safety standards for the public. It is time for environmentalists to realize, and many already have, that worker safety must be one of their primary concerns. Pollution must be stopped where it begins, at the point of production, in the workplace.

The great scandal of industrial injury, occupational disease, and workers' compensation, is that the life of an American worker is so cheap that it costs industry less to injure or kill than to create safe working conditions. The same companies that are now very conscious of consumer safety—because they have been sued successfully and lost to injured consumers—need have little or no concern for worker safety—they are immune from liability. Having fully insured its risk, industry has turned its back on its own workers.

2

Human Oxen

When workers get hurt they are expected to minimize the effects of the injury and get on with the job. Deep in the psyche of our industrial civilization is a sentiment that the ideal worker is an extension of the machine he or she is working on. The ultimate betrayal to the employer, to his fellow workers, and indeed to himself, is to not be able to function. The worker who is seriously injured on the job becomes nonfunctional, and the nonfunctional has no place where work is being done. It is either repaired, if that is economically worth while, or thrown out. The worker who is "repaired," restored to health and ability to function as before, is the lucky one. The rest must learn to deal with their new membership in the caste of the unemployable.

The adjustment to this caste involves many levels of the worker's being. First is the adjustment to financial helplessness. The disabled worker wakes up one morning with the cold realization that not only is there no money for himself or herself and the family; there is no way to earn any. The body which is the worker's capital and source of sustenance simply can't get up and do the job any longer. No amount of willpower will suffice. There's just no way.

Spreading slowly out of this awareness of financial helplessness is the emerging new identity of the *disabled*. This is no longer a worker. This nonfunctioning thing is not a complete being in the industrial sense. He has needs as ever, in fact more needs than before. But he has nothing to give in exchange. He is already used up. The disabled worker's task is to learn to live with this new identity, to learn to wait for others to meet his or her needs in their own time on their own terms. For most, the safety net of workers' compensation turns out to be full of holes.

There are many stations in the psychological and social descent of the injured worker, and attitudes that correspond to each station. The beginning is marked by confusion, the middle stage is often hopeful, and the final, terminal attitudes are often bitter and cynical and angry. Stripped of the ability to work, to function, a man or woman in our society is bereft of independence and a sense of self-worth.

Learning to Live with It

Traumatic injury on the job happens with monotonous regularity. From the fields of California to the factories of New Jersey to the oil refineries of Texas, people are being maimed at work. Joan Thomas was twenty years old in 1974, when she got a seasonal job at a small company that processed corn seed in her hometown in central Iowa. One of her jobs was to clean out a conveyor belt. The conveyor collected corn between the belt and the rollers beneath it. Joan's supervisor explained to her how to clear the conveyor by inserting a twelve-inch-long brush between the belt and the rollers. He warned her that the task had to be done with a partner who could turn off the conveyor belt if "anything happened." The conveyor had no guard surrounding it. It ran down the center of a long room whose sides were lined with large machines that dried corn seed.

One day Joan was cleaning the belt the way she had been told to do it. Her partner that day was her mother-in-law, whom I'll call Martha. Martha took a short break to go to the bathroom. With her partner gone, Joan was the only person in the room. The drying machines were making their usual din and the belt was clacking away. Despite the warning that she should not clean the conveyor while alone, Joan knew that the informal practice at the plant was to continue to clean it; she really couldn't stop just because her partner was gone for a few minutes. Stopping would have seemed like malingering. Joan accidentally turned the brush the wrong way and it was drawn into the conveyor belt, between the belt itself and the rollers underneath. Her arm was drawn in also. She tried to pull her arm out but the effort was useless: she wasn't strong enough to fight the inexorable pulling force. Fortunately, there was a guardrail on the other side of the belt. Joan's arm was drawn in almost to the shoulder, but the rest of her body was spared. Joan's screams were drowned out by the noise of the machinery. When they found her she was no longer screaming. They cut the belt and took apart the machinery under it in order to get her out. Joan remembers smiling and asking why they were carrying her out. It wasn't her legs that hurt, she explained to the people helping her.

Joan's main occupation for the following few years was simply recuperation. The skin on her right arm was damaged as in a serious fire. Not much was left.

She was lucky to have gotten good medical treatment. Many workers fare much worse. First there were the debridement operations. Debridement is defined by the dictionary as the "surgical removal of lacerated, macerated, or contaminated tissue." The flesh on the inside of Joan's arm had been ground up with the grease and dirt of the rollers. Much of it and the skin covering it was useless. It took three operations under general anesthesia to clean the wound enough so that skin grafts could be made. Just before each of these operations the doctors were optimistic enough to think that the grafts would be possible right away. They warned Joan that when she came out of the anesthesia her legs would hurt, since the skin to be grafted was to be removed from her thighs. Each time she woke up she was disappointed that her legs didn't hurt. This meant that she would have to undergo yet another operation. Finally, a little more than a month after the injury, in the fourth operation, the grafts were performed.

The skin grafts on her arm were successful at first, but there were problems with her hand. The transplanted skin over a tendon in the center of her hand didn't take and the two portions of new skin on either side of the unsuccessful patch grew together. This drew the opposite sides of her hand toward each other. The little finger and the ring finger were pulled toward the middle of the hand and twisted. Joan's right hand became a claw.

Joan was injured on October 29, 1974. In January of 1975, after four operations, she was referred to a surgeon at the University of Iowa for the problem with the hand. In Iowa City the doctor did a split-thickness graft. He removed all of the skin that had been previously grafted on Joan's hand. A rectangular section of skin on her abdomen approximately six inches by four inches was cut on three sides but left attached on one side. The flap of skin was attached to her hand on three sides while still attached to the belly. In the same operation a patch of skin was removed from Joan's leg and sewn onto her abdomen to replace the skin that was to become part of her hand. Joan's hand remained attached to her body for three weeks, until the doctor was satisfied that the graft had taken.

While her right hand was being treated, the grafts on Joan's arm contracted and became too tight. She found that she was unable to straighten her arm. Four more operations were done to loosen up the grafts to allow Joan to move her arm and fingers freely. Joan had a total of nine operations over two years.

Nine years after the injury, Joan Thomas has recovered as much as she's ever going to recover. She has a long scar from about three inches below her armpit down to the middle of her right hand. The skin on the inside of her arm is textured instead of smooth, and the flesh beneath it looks mauled. The skin on the palm of her right hand is smooth, without the lines and wrinkles you expect to see on the palm of a hand. The worst scar is the one on her abdomen. Joan has scars all over her legs from where the skin was removed to provide the grafts for her arm.

She can use her arm, although it often hurts. She types and writes with her injured hand, and she has as much strength in her right arm as her left. Sensation in the right arm is different than in the left. The injured right arm and hand are extremely sensitive to heat and cold. The little finger and the ring finger are often numb.

Joan's injury and its aftermath have changed her life. The summer of the accident she was married to a man who was overseas in the army. He came back to Iowa on a hardship leave after Joan was hurt, and was sympathetic. Nevertheless, the marriage couldn't stand the shock of Joan's injury and the condition in which it left her. After the divorce, Joan left Iowa and moved to California. In 1978 she met a man who was not put off by her scars. They fell in love and were married in 1979. Joan is now a full-time student at the University of California, majoring in psychology.

Joan Thomas lives with her husband in an upper flat in an old-fashioned Victorian house in Alameda, California, a relatively quiet backwater in the busy San Francisco Bay Area. She's a slim woman in her late twenties. Wearing shorts and a short-sleeved blouse, she looked normal—until you glanced at her arm. Or her legs. She's not the type to wallow in self-pity or anger. Still, beneath her stoic Midwestern exterior, the feelings run deep. On being asked what she's lost as a result of her injury, she says: "I don't know what to say. I've been angry many times. It's just something you have to learn to live with. It's going to always be there [sobbing suddenly]. I'm a young woman. I was twenty years old when this happened. I can't help but feel self-conscious when I wear short-sleeved blouses. When I'm around friends it doesn't bother me too much. It doesn't seem fair; it doesn't seem right. People get compensated for much more trivial things than this."

About a year and a half after the accident Joan contacted a lawyer to make sure her rights were protected. He didn't seem to know what he was doing, and she soon fired him. She was then approached by the lawyer who represented the workers' compensation insurance carrier that insured her employer. This lawyer suggested that she sue the manufacturer of the conveyor belt that had hurt her. Joan readily agreed, thinking that this must be the right way to get compensation for her injury.

A lawsuit against a manufacturer that is not the employer of the worker who has been injured is called a "third-party suit." It is one possible legal remedy for a worker who has been injured on the job. Workers' compensation laws bar the worker from suing an employer, but in many states they allow such third-party suits. If there is a judgment or a settlement in favor of the worker, the workers' compensation insurance carrier is entitled to recover the amount of money it spent to help the worker, including the money it spent on medical bills. So it is no wonder that the lawyer for the carrier was eager for Joan to sue.

But it was not a clear case of negligence, and third-party suits require this. The conveyor belt had been sold in pieces to Joan's employer ten years before the accident and had been assembled improperly by Joan's employer. Five years after the accident the case finally came to trial and Joan lost. She got nothing from the manufacturer. A year before the verdict in the third-party case, Joan had signed a form releasing the compensation carrier from any further liability for her on-the-job injury.

Joan received $70 per week for two and a half years and her medical bills were paid for. When the insurance carrier cut off payments she went back to work, although the arm was still hurting. She managed somehow. The insurance company saved itself a lot of money on what might have been a troublesome claim. The biggest winner was Joan's employer, which was not affected financially in the least by Joan's injury.

There is a bitter postscript to this story. In 1980 another woman suffered an almost identical injury on the same conveyor belt.

A Faithful Employee

Although industrial injuries occur with greater frequency among inexperienced workers, length of service on the job is no guarantee of safety. Some workplaces are so perilous that the workers are virtual front-line soldiers in a war for more profitable production. The coke oven department in many steel mills is a little like a war zone. Not many people work in the coke oven by choice. It's the first job you get if you want to be a steelworker. At a Bethlehem Steel plant in northern Indiana, if you're hired off the street you have to work in the coke oven for a year before you can transfer to another department. Dorothy Hanna is one of the few who chose to stay. She had worked there for two and a half years when she finally was injured.

A piece of metal fell from three or four stories above Dorothy and hit her a glancing blow. It was a miracle that she wasn't killed or crippled. Dorothy is a large, strong woman. Immediately after being struck she recovered from the blow enough to be aware of people and events around her. Dorothy's clothes were ripped, and her foreman threw a jacket over her. The piece of metal had cut through her three sets of flannel underwear, her T-shirt, her flannel shirt, and her yellow protective suit. All this clothing had to be worn to protect her from the extreme heat of the battery of ovens. The metal had sliced her bra off.

Paramedics came to take her down from the platform where she had been struck. But Dorothy was alert enough to refuse to be removed by them, down the two steep stairways to the plant floor. All she could think of was an incident she had seen a few months before. A man had been close to an explosion and was burned all over his body. When Dorothy saw him he wasn't quite dead yet. The paramedics who were carrying the man down to the floor dropped him the last several feet. When they dropped him he made moaning sounds, although his tongue was burned out. Dorothy wouldn't let the paramedics touch her. So they brought the boom crane in to take her down. She was put in an ambulance and taken to the company doctor. I'll call him Dr. Reynolds.

Dorothy was given a chest X ray and allowed to rest for an hour and a half in the company clinic. Dr. Reynolds then told her to go home. In retrospect she understands that she should have been sent to a hospital, but at the time she just did what she was told. "Everything was fuzzy, so like a dutiful soldier I'm doing my dutiful duty. Tell me to stand up and I go. And they tell me to turn and I turn. Now I know it's ridiculous. It's stupidity on my part. But then I didn't know any better, see. And I feel that like the company'd take care of it, the company knows, they say you're all right, you're all right."

Dorothy attempted to drive home in her own car, but had to stop near the mill because she was seeing double and feeling very nauseated. She managed to get someone at a liquor store to call her husband, who came and drove her the thirty miles to their house. After she got home she found out that the company expected her to show up for work the next day. In the morning her supervisor told her to come in to work so the company doctor could look at her. She said that she couldn't move. "Then they called back and told me to take a couple of aspirins." Dorothy stayed in bed for a few days, and when she felt well enough to move she went to see her family doctor. Dorothy's doctor allowed her to stay out of the hospital, since the worst appeared to be over. She needed frequent office visits, however. Since the family doctor was uncomfortably far away from her house, Dorothy agreed to go to someone the company recommended, a Dr. Smith. If the company wished, it could have made Dorothy go to Dr. Smith even if she hadn't wanted him to treat her. In Indiana, an injured worker doesn't have the right to choose his or her own treating doctor; the company has that right. The employer has no obligation to pay a doctor it hasn't chosen.

Dorothy's medical condition is still unclear. She did suffer a concussion. One of her eyes is now off-center and her vision is slightly distorted. She has had a great deal of physical therapy, and is beginning to be able to walk normally. Being a tough and determined woman, she has been working with weights to regain her strength so that she can return to her job. Before the accident she had wanted to become a crane operator and had taken the rigorous eye examination required for the job. She passed with flying colors. In particular, her depth perception was excellent. All that is gone now.

Her shoulder often feels numb; it may be numb for a whole day at a time if she has slept on it the night before.

While she was unable to work, Dorothy received workers' compensation temporary disability payments. She got the then-current maximum in Indiana, which was $130 per week. But in order to continue to get workers' compensation, she had to stay on the good side of Bethlehem Steel, and it wasn't easy. She was required to drive the sixty-mile round trip to the mill every two weeks to see Dr. Reynolds. The visits were gratuitous, since she was being treated by another doctor who reported to the company on a regular basis. Despite her pains, Dorothy would dutifully report to the company doctor: "He'd sit down and he'd say, 'How's it going, Dorothy?' And I'd say, 'All right.' And he says, 'How often are you going to the doctor?' and I'd say, 'Twice a week.' He says, 'Anything bothering you? And I'd say, 'Oh, just the usual stuff.'" But if Dorothy lost her temper, something she's been known to do now and then, or if Dr. Reynolds was in a bad mood, she'd be required to come in to the company clinic every week for a while instead of every two weeks.

Although the visits to Dr. Reynolds were humiliating and useless, Dorothy had to make them. If she didn't, her $130 per week workers' compensation temporary disability payments would be stopped. On the day of one scheduled visit, Dorothy had the flu. She called the workers' compensation man at the mill office, someone named Rick.

"I told him I had the flu, and he told me I'd better be in. He'd say, 'You better beee here,' in a singsong voice. And I'd say, 'Rick, I'm sick.' He'd just say, 'You better beee heeere.' And I told him, 'I am going to the doctor—why do I have to come out to the plant too?' And he kept telling me that it was something to do with workman's comp. That you have to. 'Paperwork, Dorothy, paperwork. Has to do with paperwork.'"

Dorothy missed that useless appointment with the company doctor and didn't get a compensation check until she was able to go to the plant and see him again.

Dorothy was off work for almost a year and still recuperating from her on-the-job injury when she was called in to the mill for a special appointment with Dr. Reynolds. The company doctor performed a perfunctory physical exam and told her to go back to work, or say that her condition was not work-related (hence not compensable under workers' compensation). As Dorothy relates the incident, the company doctor was not very subtle:

"He tapped my wrist, he tapped my two knees, and he signed a form—it was already written out. He says, 'You're released to go to work, Dorothy. As far as Bethlehem Steel is concerned, your injury that you sustained here, you are now well of. Now if you have any more problems you go over to the insurance department and sign up for SIP payments. And of course you'll get more money, you know that, Dorothy.'"

SIP means Sick Insurance Premiums, disability insurance covering Bethlehem Steel employees for non-work-related injuries. The SIP payments would have been $180 per week. But in order to get the SIP payments, Dorothy would have had to sign forms declaring that her injuries and present condition were not work-related. If she did not sign the papers, however, she would have no income at all, since Dr. Reynolds had declared that her work-related disability was now over. This despite the opinion of Dorothy's treating physician, Dr. Smith, who said that she could not go back to work because her injuries from the on-the-job accident had not yet healed.

Dorothy did not sign the papers. Instead she walked out of the company clinic and went to a local restaurant and called the attorney whom she had hired to represent her for her workers' compensation case. He told her not to sign anything. Fortunately, Dorothy's husband is a well-paid skilled worker, and although there is real financial strain, she will get by without workers' compensation or the SIP.

More than anything Dorothy wanted to return to her job. But she knew that she wasn't ready for the extremely arduous work. She was afraid that if she returned to work before she was ready, she'd reinjure herself.

Since Dorothy would not apply for SIP, the company doctor said that she was able to go back to work. He also said that her disability was no longer due to her job-related accident. In other words, she wasn't disabled, and if she was disabled it wasn't job-related. This type of reasoning is called by lawyers "pleading in the alternative." It is a well-known and acceptable practice in the legal profession. After all, it is the judge or the jury that has the responsibility of deciding which story to believe. Dr. Reynolds' medical opinions smacked more of the courtroom than of a professional dedicated to healing. Dorothy's disability payments were cut off and her lawyer filed for a hearing with the Indiana Industrial Board. The first hearing, set for several months after the benefits were ended, was postponed another few months at the request of Bethlehem Steel's lawyers.

Dorothy is confused about events since the accident. She was an extremely loyal, even devoted employee of Bethlehem Steel. She doesn't understand why her workers' compensation was cut off and why she has been hassled so much. When she finally hired a lawyer, she could scarcely believe what he told her about her "rights," or rather lack of rights, under the workers' compensation laws of Indiana. Her lawyer, David Hollenbeck, told me that at their first meeting Dorothy ended up screaming at him and pounding on his desk. As Dorothy said: "Oh my God, I was shocked. I thought we had rights. That's all I kept telling Dave, I said, 'My God, we've got rights. I mean, they're violating my rights.'"

Dorothy couldn't understand how the company could cut off workers' compensation benefits to someone who had been so good: "I'm doing everything I'm supposed to do, everything. I'm not skipping doctor's appointments, I'm not skipping therapy, which I would love to do. I'm getting calluses running back and forth and I'm doing everything where they can't say, 'Dorothy, you didn't do this or, Dorothy, you didn't do that.' It was incredible, just incredible, that this could happen to me, because I gave the company . . . if I went to work there eight hours, I did twelve hours' work. They shouldn't have done it. I can't get it together in my head. I'm such a faithful employee. I've never been late to work, ever, are you kidding? If this was me coming in the door, me seeing me coming in, I'd hire me in a second, man. Everybody needs somebody like me. They do. Hey, man, I'll protect your equipment, I'll work like hell for you. All the foremen know me, everybody knows Dorothy. I don't hide from work. So I figure, this isn't gonna happen to *me*. They did me wrong. Dammit, they did me wrong, and I was doing them right all this time."

The injury has made Dorothy assess her life. She often dreams about the accident, and once in a while she'll lose herself in a kind of reverie, words repeating themselves in her mind: "Oh my God, I could have been killed, oh my God, I could be dead, oh my God, I could have been killed." Sometimes she thinks that she might have been totally paralyzed instead of being killed. This is the worst thought for her—she's sure she'd rather be dead. Dorothy doesn't know what the future will bring her; she's trying to cope with the present.

A loud, hard-living woman, not afraid to tell anyone to go to hell if she thinks they deserve it, she says in a voice as soft as a child's, "I'm thankful I'm walking around." She says, "I want to get out from under the mess I'm in, see. Being that we made some money, we're gonna have a nice retirement and help the kids out—you know the interest rates the way they are now. These kids aren't gonna be able to afford homes and stuff like that. It's always been my husband and I. We're very close, my husband and I are, and our circle is our family and we don't expand that circle very much. And I could have lost it very easily."

Dorothy has learned to appreciate some other aspects of life and to get satisfaction from them. "I love antiques. I like to preserve old things. I hate to see old homes destroyed. I hate to see old buildings go down for a parking lot. I want to save our library in town; things like that are important to me. Before it was go to work, work a double—I worked ninety-six hours out there one week. Before, my husband knows I overdid it before. He used to say, 'The mill, the mill, the mill, that's all I hear is the mill.' And it's still the mill, the mill, but now I'm angry because they do me wrong."

The chronology of Dorothy Hanna's workers' compensation case says a lot about the stately pace of the system in Indiana, which is not unlike most other states.

When Dorothy's workers' compensation benefits were cut off, her lawyer filed for a hearing before the Industrial Board of Indiana. The hearing was scheduled for March of 1981. It was delayed until June at the request of Bethlehem Steel's attorneys. In August, the board ruled in Dorothy's favor. Bethlehem Steel appealed that ruling and was allowed to postpone payment of her workers' compensation until the appeal was heard and decided. The full Industrial Board ruled in Dorothy's favor on December 17, 1981. Bethlehem Steel had the right to appeal that ruling but decided not to.

Dorothy is now recovered as much as she will recover from the injury. She has a permanent partial disability, which her treating doctor says reduces her ability to work by 30%. Bethlehem Steel will not accept this percentage and refuses to agree to pay Dorothy anything. Dave Hollenbeck, Dorothy's lawyer, will have to go through the whole process of litigation once more. He will request another hearing before the Industrial Board, the first in another series of hearings.

A Classic Case

Of all on-the-job traumatic injuries, back injuries are probably the most common cause of disability and workers' compensation litigation. Legions of orthopedists and armies of lawyers (both defense and plaintiff) owe their livelihood to the vulnerability of the human back. Unlike most other traumatic injuries, there is often a serious problem of proof that the injury was caused by the job. There are several major reasons for this proof problem. First, symptoms do not always appear in their most virulent form at the time of injury. It is common for a worker to injure his or her back one day and barely feel it, waking up with horrible pain the next morning or even weeks later. Secondly, cumulative injury to the back at work mimics degenerative changes that are often the consequence of the aging process.

There are millions of jobs that strain the human back. Because the straining is so constant and so taken for granted by everyone at work, a worker whose back is injured may not even think of the injury as compensable under workers' compensation. Indeed, in many states an injury is compensable under workers' compensation only if it is the result of an "accident." In these states, a back injured as the result of years of work at the same arduous job is not considered a work-related "injury." The more progressive states have accepted the concept of "cumulative trauma"—the slow accumulation of small injuries to the back eventually causing disability.

The anatomical complexity of the human back and the variety of all its possible ailments makes it an impossible subject for adequate coverage under the limited categories of the workers' compensation laws. Injuries must be proven to be work-related and the degree of disability must be established by "objective" evidence. Many workers with bad backs fall afoul of these artificial legal requirements. Whether a worker gets compensation for a bad back usually depends on the competence of the doctor who is treating him and his ability and willingness to diagnose a work-related back injury. The effect of a bad back on the life of a worker who has always depended on his or her strength to make a living is usually devastating. When Mike Sarkis injured his back while working as a warehouseman in Philadelphia in 1962 he didn't know anything about workers' comp. In his early twenties, he thought that he was completely on his own. No one ever told him that workers' compensation might help. So he quit his job at the warehouse and went to see his family doctor, who told Mike that he might have "an arthritic problem." The doctor referred Mike to an orthopedist, who concluded that Mike's problem was psychosomatic. Mike's life became a hell:

"I went off and for the next three years was in agony. I slept a maximum of three to three and a half hours a night before I'd wake up crying from the pain. It would just tighten up so much, and then gradually the whole sciatic nerve problem got worse and worse, to where I couldn't walk on the left side. When I had to step on my left side I had to grit my teeth because the pain was so bad."

Mike moved to California and found a doctor who diagnosed his problem as a herniated disc. He was operated on in early 1966, and immediately felt better. Later in the year, however, the pain began to come back. Another operation was performed, and this one seemed completely successful. Mike's pain was relieved and he was able to lead a normal life for the first time in almost four years. Normal life meant hard manual labor. For the next ten years he worked as a rigger in an oil refinery, a welder's helper working outdoors on twelve-hour shifts, and as a warehouseman.

Mike had always earned his living with his strong back. In the late 1970s Mike was living in the San Francisco Bay Area, working out of the Teamsters' Union Hall in Oakland. He did a lot of very heavy work, the kind of repetitive, mindless lifting and carrying that many middle-class people assume is now automated. Mike unloaded railcars full of rice, for example. The sacks of rice weighed fifty or one hundred pounds, and Mike and his fellow workers would pick them up and put them down every few seconds for eight hours a day, five days a week.

The insult to Mike's back that finally put an end to his career as a manual laborer occurred when he was working at a regular job at a warehouse in Hayward, a town just south of Oakland. In late March of 1978, Mike was throwing a seventy-five-pound wooden pallet to the top of a stack of other pallets. The stack stood between two metal posts. The pallet Mike was throwing hit the edge of one of the posts and fell back on him and hurt his back. Mike was taken to the industrial doctor, who diagnosed a low back sprain and sent Mike home to rest for a few weeks. The doctor did a perfunctory exam, never talking to Mike and never getting an accurate history of the type of work Mike had been doing when he was injured. Mike made a point of telling him of the earlier back operations, however.

Mike received workers' compensation temporary disability payments of $154 per week, the maximum in California at the time, for three months. Liberty Mutual, the largest workers' compensation carrier in the country, then sent Mike to a different doctor, who examined him and said that the condition of his back was "permanent and stationary." This means, in California law, that an injured worker can no longer collect "temporary" disability payments. Mike's workers' compensation was stopped pending final resolution of his case.

The next year and a half was a nightmare for Mike Sarkis. He had no other resources besides his back. He was luckier than many men his age, thirty-seven. He was single and didn't have small children to support. For a while he was able to get some money from California's State Disability Program (SDI), which provided the same benefits as workers' compensation for up to twenty-six weeks. Unlike workers' compensation, state disability eligibility requires only proof of inability to work, not that the injury is job-related. Also, it makes no distinction between "temporary" and "permanent" injuries.

Although he had a college degree, Mike had never used it. Eighteen years after graduation it was probably useless. The only work he knew was manual. When the SDI stopped he went back to the warehouse, hoping that he'd be able to do the job. For a while he was lucky, getting orders to fill that weren't really heavy. But then the Christmas rush came:

"All of a sudden all the orders were heavy, and I just went downhill fast. Because I mean you're talking about pushing real heavy carts with big heavy bulky stuff. Pretty soon I was in a lot of pain, and once it starts to go it just gets worse and worse. And we're talking about my hip and my leg and my whole sciatic nerve, so that when I go to step on it it's like stepping on a spike." Mike continued to work despite the pain. "I just toughed it out. I was in debt and whatever I made there had to last me. So I just was into staying as many days as I could. The two times I went back I did that kind of macho thing where I thought I was going to be able to control the pain and make sure I didn't hurt myself. I was stupid and I pushed it.

"I lasted from November 14 to January 19, and the only reason I lasted that long is that I had Thanksgiving holidays and Christmas and I got time off. I tried then not to even work five days a week, I tried to really space it out, drag it out." Finally, Mike had to stop working. "It just got real clear. I knew the week of January 19 [1979] that I couldn't do it anymore. I would just try to make it to the end of the day, and it became obvious that I just couldn't go any further. I could not walk without gritting my teeth. I limped around as long as I could—I was just trying to get in as many hours as possible. And then, finally I had to draw the line because the pain was too bad and I felt that it was stupid to push it any further."

Mike's goal was not only to make enough money to live for a while, it was also to try to continue his health insurance. When he finally had to quit he managed to console himself a little:

"So I was angry and hurting an awful lot and dejected, but I knew I had actually accomplished a lot because the first goal when I went back was to work fifteen days, which is a requirement that you have to work in order to sign up again for state disability insurance. Their requirement is fifteen days without any time off for your injury, but really it's without any time off. You got to show that you can do the job without any restrictions. So that was my goal, to work at least that long, longer if I could, so that when I got off I would have reestablished my health and welfare, which they had to pay me for six months when I went off because of my back. The union would pick up three and the company would pick up three so I'd be OK for a while, plus I had some money plus I could go on state disability. Which means I could exist."

Mike stayed off work from that time for six months. When he exhausted his state disability payments and had no money to pay the rent, he went back to work at the warehouse. He told his supervisor that he couldn't do the work but that he had to come back. He desperately needed the money. The supervisor didn't say much. He let Mike know that it was his, Mike's, problem. Mike tried his best to work, but found that "It was the same routine, gradual deterioration, finally coming up to a point where I could barely walk." On October 2, Mike quit his job so that he wouldn't have the temptation of returning to work. "This time I knew that I couldn't go back again without really risking crippling myself. I was afraid that I was developing something new that was just going to continue to deteriorate to the point possibly where resting would not let me recoup the damage . . ."

Although Mike's back injuries were attributable to work, and as a result of them he could not work, his workers' compensation was stopped only a few months after the injury in 1978. So he hired a lawyer. The lawyer was recommended by his union. Mike never laid eyes on this lawyer, although he did once talk to him on the telephone. Mike eventually went through four lawyers, none of whom did him much good.

Mike has seen many doctors, not so much for treatment as for litigation purposes. First the company doctor, then the insurance carrier doctor, then a doctor chosen by Mike's lawyer, and finally a doctor chosen by agreement of the insurance carrier and Mike's lawyer, the "Agreed Medical Examiner." Because the case dragged on for so long, Mike had to go to each of these doctors twice.

Since his workers' compensation was cut off soon after his accident, Mike has fought to get it restored. Under California law he was entitled to receive vocational rehabilitation benefits if he couldn't go back to his job. After great struggle, he managed to get Liberty Mutual to pay him these benefits. Disability payments were resumed in September of 1980. The state of California is paying for the expenses of a one-year program of paralegal studies.

The pain, the doctors, the lawyers, the state bureaucracy, the insurance company bureaucracy, the worry about money, the dread of the future—this witches' brew has been the substance of Mike's life since he was hurt on the job. He's not the same person he was three years ago. He says, "I feel that it's changed me permanently. My sense of myself, the pressure of dealing with the attacks on my person, the physical pain, the falling into the mistake about being expectant, the hope that something's going to happen, you're going to get a good report, or maybe if I do this right, or God, you got this, another doctor's appointment. Everything's on the line. Money. For two years I never went out, never did anything that cost money. I stopped a relationship with a woman that I was with, and then I never really have been able to get anything going since. I've drastically changed about money because being ten thousand dollars in debt owed to a friend who has cancer, who is dying and needed the money, while all of this stuff was going on and being subject to the insurance company and the state, where they could play with you so easily. I was racked emotionally."

Mike is now enrolled in a paralegal training program at a local junior college. He is also doing an internship with a law firm that specializes in workers' compensation litigation. Mike attempted for six months to help other injured workers cope with the intricacies of California's workers' compensation system.

Wasting Workers

The conditions that cause bad backs, carpal tunnel syndrome, mangled arms, broken bodies, and broken lives do not change. Despite the regularity of injury, the cost to workers, to employers, and to society, there have been no changes where Joan Thomas, Dorothy Hanna, and Mike Sarkis worked. These people and millions like them are maimed with a nonchalance that a more humane society than our own would find astonishing.

Experts in industrial injury and participants in the workers' compensation system know all about the routine injuries and the jobs that cause them. Few of these professionals—at least, few of the people who manage workers' compensation or who represent workers or employers—have ever considered the possibility that the jobs that cause the injuries should be restructured to prevent injury. Once in a while, a union presses demands for safer workplaces, usually after conditions have gotten so bad that workers are actually being killed on the job. But the routine injuries go on and on.

Government regulation of workplace safety through the Occupational Safety and Health Administration (OSHA) has improved conditions somewhat, but government inspection and regulation will never be enough. Workers must advocate for themselves. They must insist on safe workplaces. Unions that have focused their attention on safety problems should be commended and urged to do even more. Unions that have done little should be shaken up by the rank and file, whose own bodies are on the line.

3

Occupational Disease—An Epidemic

The problem of safety hazards on the job is chronic; the problem of health hazards that cause occupational disease is fast becoming acute. It is not an exaggeration to say that the lives of millions of American workers are in jeopardy because of exposure to toxic substances. The future for these potential victims is bleak. The workers' compensation system that offers so little to victims of traumatic injury is practically inaccessible to those suffering from an occupationally induced disease. There will be great numbers of these victims in the years to come, because occupational disease is fast becoming epidemic.

In the first two decades of this century workers by the thousands were falling to their deaths from unprotected catwalks, having molten steel poured on them, and being run over by a variety of heavy machines.

Newspapers of the period carried alarming stories of mutilation and death in factories and mines. It was the heyday of muckraking journalism, and there was no dearth of sensational stories in the American workplace. Upton Sinclair eloquently exposed the hell of the slaughterhouses in his famous novel *The Jungle*. That amorphous but inexorable force, public opinion, began to demand reform.

Attempts at reform focused on the acute problems of traumatic injury rather than the long-term problems of occupational disease. It was the traumatic injury, the sudden snap of an arm or a neck, the brilliant flash of a human body consumed by liquid fire, that had captured the imagination of the public and quickened its indignation at the harsh working conditions of so many Americans. It was this sense of injustice that state lawmakers sought to appease with the first workers' compensation laws and the first halting attempts at regulation of workplace safety.

Occupational disease presents special problems, different from those of occupational injury. Although diseases of occupation have been known for centuries—the scrotal cancer of chimney sweeps and the nerve damage of lead workers, for example—early workers' compensation laws and early government safety-regulators were not much concerned with them.

With its long latency periods and invisible hazards, occupational disease was much less dramatic and seemed at the time a much less immediate problem, something that could be tackled in the future.

More than seventy years after the enactment of the first workers' compensation laws, the problem of occupational illness has still not been tackled. Of the 581,000 Americans severely disabled due to occupational disease, no more than 5% receive workers' compensation.

One of the universal principles of workers' compensation is that for a disability to be compensated, it must be caused by work. Work-relatedness must be proved. It is obviously easier to prove that a foot was smashed by a machine at work than it is to prove that one year of exposure to asbestos long ago caused a recent case of lung cancer. Procedural difficulties are extraordinary. The burden of proof is on the worker, who must prove that his or her disability was caused by conditions at work. He must prove that he was in fact exposed to a particular substance, that the substance was harmful, that it harmed him, and that his disability was caused by this harm and not by some other, nonindustrial cause.

The same workers' compensation laws that make it almost impossible to recover for occupational disease in some states and very difficult in others take away the employees' right to sue for damages in court. Workers' compensation is the "exclusive remedy" for workers—even when it is no legal remedy at all.

Original workers' compensation laws in many states simply excluded occupational disease from coverage. For example, North Carolina, one of the centers of the textile industry in the United States, did not compensate cotton-mill workers for cotton-dust disease, byssinosis, until 1963.

Workers' compensation laws that excluded occupational disease or made it almost impossible to prove sent a message to industry that the government and the public were not interested. There was no economic penalty to employers who continued to expose their workers to dangerous substances. Since the raison d'être of business is the making of a profit, it followed logically that industry would expose workers to occupational hazards if it was profitable to do so. Nowhere was it more profitable than in the asbestos industry.

An Asbestos Worker

John Vandyne had the misfortune of working for fifteen years in a California factory that used asbestos to manufacture insulation for pipes. The asbestos would arrive at the plant in boxcars, a loose powder packed in loosely woven burlap bags weighing fifty to one hundred pounds. The bags would be unloaded by hand and the contents dumped on a wide conveyor belt. There was enough asbestos dust in the air most of the time so that the workers in this factory were covered with it at the end of the day; their clothes were stiff with it.

Vandyne certainly did not consider himself unlucky at the time—quite the opposite. He was earning good wages and working for a large, solid company, Johns Manville Corp. Good money, job security; he thought he had it made.

No one at the plant thought that asbestos dust was harmful. One supervisor even said that it was good for you. He claimed that it helped your digestion. The manager of the plant, August Schilling, was a kind of father figure to Vandyne. Like Vandyne, Schilling grew up in a strictly conservative Dutch family and was an embodiment of the virtues of hard work, reliability, and self-discipline. Schilling also had the distinction of being a real veteran in the asbestos industry. He had started working around asbestos in 1930.

By 1968 Schilling was in his late sixties and unable to walk around the plant; he couldn't catch his breath if he walked more than a few steps. So he had a little electric cart with a bicycle type flag on it that he drove from one section of the plant to another. Of course, no one knew why Schilling was short of breath; the workers thought that he was just getting on in years. Schilling spent a fair amount of time talking to Vandyne about the factory, about Johns Manville, a company which had taken care of him since he was a young man, and about asbestos, a naturally occurring mineral with over 3000 industrial uses.

Schilling's breathing became worse and worse. One day some Johns Manville executives flew in from corporate headquarters in Manville, New Jersey. Vandyne knew in advance they were coming because the plant got a thorough cleaning—something that invariably signaled a visit from the home office. The cleaning involved use of powerful blowers to blow the dust into large piles, which would then be removed. Like a proper housecleaning, the blowing would start from the top and work down. The whole process took forty-eight hours, and for a while the factory was so filled with asbestos dust that you couldn't see for more than three feet in front of you. Vandyne, who was often one of the cleaners, remembers how after the cleaning was over his eyes stung for two days. The dust would be embedded in his nose and ears.

The plant was cleaned; the home office people came and took Schilling out to lunch. We may never know what was said at that lunch, but it is evident that Schilling was an embarrassingly visible reminder of the dangers of asbestos, a danger that Johns Manville did not admit at that time. According to Vandyne, the home office no longer wanted Schilling at the plant with his wheezing and his electric cart. He was unceremoniously retired—he never returned from lunch that day. Vandyne heard several months later that Schilling had died. The new manager of the plant said it was emphysema.

In what the military used to call a "surgical strike," Johns Manville removed Schilling, who was a bad example of asbestosis. The company retained working procedures that exposed employees to huge amounts of asbestos dust.

John Vandyne, once a strapping young man able to meet any

physical challenge thrown his way, has followed at least one aspect of his mentor's career. He's got a bad case of incurable asbestosis.

At age fifty-six, Vandyne sits a lot. To walk more than a few steps is painful and exhausting. The feeling he gets when he overexerts himself (and a man like Vandyne finds it hard not to overexert himself) is a lot like the feeling of drowning very slowly, almost being able to catch your breath, but not quite.

One would think that workers' compensation would at least lift a little of the financial burden from Vandyne's weakened shoulders. His employer, Johns Manville Corporation, the largest asbestos goods manufacturer in the world, has been fighting his claim for four years. As of this writing the case has not been settled. John Vandyne has gotten no compensation from the time he stopped working at age fifty-two.

Fortunately, there exists a Social Security disability program which gives Vandyne $450 per month. Also, his wife works, so they are able to survive. They have had to move out of California's hot, dusty Central Valley, where John's family and his wife's family live, to the cool clean air of the Monterey coast. Although they know no one in the new town, at least John can breath a little better. Deciding to move was painful but inevitable; breathing has had to become a number one priority. Luckily, John's wife didn't resent him for it and their marriage is intact.

John Vandyne's case is a typical story of occupational disease in the United States. If anything, John is more fortunate than many of the half million of his fellow citizens who are totally disabled due to damage done to them by conditions on the job. At least he knows what he's got.

Asbestos-related disease is one of the best-known occupational diseases, because of the great numbers of people afflicted with it, the fact that asbestos has been proven to cause cancer, and the media attention to the problem. Asbestos can cause lung cancer, usually fatal, and a cancer called mesothelioma, invariably fatal. These cancers may develop thirty or more years after exposure to asbestos. Vandyne has a noncancerous restrictive lung disease caused by asbestos; it is called asbestosis. He doesn't know what's in store for him. The asbestosis will probably get worse; he may also develop one of the cancers. John is among millions who have been exposed to asbestos; he is not yet among the many thousands of Americans who are dying because of that exposure.

According to a recent U.S. Department of Labor study, 50,000 people per year die from asbestos exposure in the United States. They die from mesothelioma, from lung cancer, from asbestosis and its complications. Death from mesothelioma, a cancer of the pleural lining surrounding the lungs, is particularly painful. The cancer encases the lungs and then crushes them inward, slowly asphyxiating the victim.

Almost no one lives longer than eighteen months after the diagnosis of mesothelioma. The actor Steve McQueen, who had worked around asbestos when he was young, fought the death sentence of mesothelioma with exotic remedies in foreign clinics, to no avail. Lung cancer is also very painful and it can last a little longer, although only 5 in 100 live more than five years after diagnosis. Asbestosis can go on for years and leave the sufferer unable to get out of bed, gasping for every breath, even with an oxygen tank, just to stay alive. Death usually comes as a result of complications involving the heart.

A Cotton-Mill Worker

Another deadly occupational disease is byssinosis, or "brown lung." This disabling lung disease occurs among textile workers, one of the oldest occupations in our industrial society, and is caused by cotton dust.

Betty Smith suffers from brown lung. She is a resident of the small town of Erwin, North Carolina. Since 1900 Erwin has been the home of one of Burlington Mills' largest mills. The mill produces prodigious amounts of denim, which is made of cotton. Betty Smith worked in the mill for forty-one years before retiring due to byssinosis. She is a member of the Brown Lung Association, an organization of former textile workers in North and South Carolina who have all been disabled by byssinosis.

Erwin is a town totally dominated by its one great industry, the Mill. But it doesn't look like a mill town. It's a pleasant-looking place, verdant on a hot Carolina morning in June. Approached on

Highway 217 from Fayetteville, there is no evidence that Erwin is the site of the largest denim factory in the world. The mill itself is a surprise. The front is a building in the style of 1959 suburbia, a one-story façade of smooth, reddish brick with the words BURLINGTON ERWIN MILL, in discreetly small white letters.

Betty Smith's house, a few blocks from the mill, can also be called pleasant. It's an old white frame house; in California they'd call it a one-story Victorian. The neighboring house is a nice distance away; the other side slopes down to a stream. There are lawns around the houses but no evidence of a garden.

Betty Smith is a woman in her middle sixties, comfortably fat. Wearing a faded housedress with an old-fashioned print as she usually does, she looks like a picture-book grandmother. Her story is in many ways a classic story of occupational disease.

Mrs. Smith worked in the spinning room of the Burlington Erwin Mill for forty-one years, from 1929 to 1970. Her job was to tend machines that spun cotton into thread. There was a lot of cotton dust in the spinning room: "It was like a real fine snow, you could see it in the sunlight by the windows." It got worse after 1960, when the windows were bricked up so the plant could be air-conditioned. "After the plant was bricked up it didn't have nowhere to escape so we just inhaled all that dust."

Around 1962 Smith developed a bad cough, "like to cough myself to death." She kept on working, of course. She needed the money and she loved her job. Besides, no one ever told her that the fine white dust was slowly building up in her lungs and would one day cause her to be very, very sick.

The end of Smith's working life came when she returned to her job after a vacation in 1970. She came back and worked four nights.

"I just didn't have the breath to talk. All I could do was tend to the job. And the lady that worked side of me she said, 'Miss Smith, what's the matter with you? Since you came back you don't have anything to say.' And I said to her, I said, 'Helen, I just don't have the breath to talk.'"

She entered the hospital thinking that her heart was bad. The nurses there told her, "Mrs. Smith, don't go back to that old mill, it's done enough damage to you." Smith didn't know what they meant at the time, but she never went back to Burlington; she just couldn't do the work anymore.

Unlike Vandyne, Smith has gotten workers' compensation for her disability. After three years of litigation she got a final and complete settlement of $14,000. She paid 25% of this to her lawyer (without whom she would have gotten nothing) and gave one thousand dollars to each of her two children. She put the rest in a savings account at the local bank. She lives on $270 per month from Social Security and a few hundred dollars' interest she gets every year from the bank. There is no pension plan at Burlington. Luckily, her house is paid off. In the short Southern winter she keeps only two rooms heated.

Workers' compensation pays none of her medical expenses—it never has paid for any. Luckily, she is old enough to be eligible for Medicare. She also carries a private insurance policy to supplement Medicare; it costs $227 per year.

There were no signs at the mill warning of the health problems associated with cotton dust. Despite the fact that byssinosis has been a known occupational disease for almost a hundred years, despite the fact that Great Britain passed a byssinosis compensation act in 1940, despite the presence of a union in the mill since the early 40s, it was not known in Erwin, North Carolina, that there was a disease caused by cotton dust, that it was called byssinosis, and that it could disable and kill. After the passage of OSHA, byssinosis became a national issue and the news filtered down to Erwin through organizers of the Brown Lung Association from Durham. Now, after eighty years of exposure to cotton dust, the people of Erwin know.

An Electroplater

The problem of toxic chemicals in the workplace is fast becoming the number-one occupational-health hazard in the United States. So many chemicals are being used for so many industrial processes, with so little control over their use, that it is little wonder that more and more workers are becoming sick due to exposure. There are many possible health effects of exposure to toxic chemicals; electroplaters, for example, are subject to no less than thirteen identifiable occupational diseases.

Jim Richardson is a fifty-nine-year-old factory worker who had electroplated Schwinn bicycles. He was also required to do some of the worst, dirtiest jobs. He often had to clean out large tanks that had been used to store toxic chemicals. He sometimes got dizzy and groggy at work. Like a good soldier, he would continue to work and not complain. Then one day he blacked out at home. He began to spit blood. Jim didn't know the cause of these occurrences, and the doctors he went to just told him that he had "bronchitis." Jim began to develop skin lesions all over his body.

Jim took off time from work during the worst onslaughts of dizziness, but he went back as soon as he possibly could. He had a family to support. At work one day, the foreman told him to climb inside a particularly noxious chemical tank and clean it out. Jim refused, saying that there wasn't enough breathable air in it. The foreman told Jim that if he didn't do what he was told he'd be laid off for ten days. This was not a union shop and Jim had absolutely no recourse. "I got a family and the rent to pay and my phone and light and my transportation to and from work. So I went ahead and cleaned the tank because I couldn't afford to be off. I had four kids at home then." His only protection was a mask that didn't really stop any of the toxic fumes.

By late 1979 Jim was unable to work any longer. He just couldn't breathe if he was on his feet for any length of time. Schwinn fired him since he couldn't work. Jim's medical insurance was dependent on his working at Schwinn. He lost the insurance when he lost the job. To get treatment for his worsening condition, he was forced to go to the county hospital, in this case the sprawling Cook County Hospital. Fortunately for Jim, there is a workers' clinic at the Cook County Hospital. Its purpose is to diagnose and treat people like Jim, workers who have been made sick by toxic substances. There are very few such clinics around the country. Jim was lucky to have happened on one.

Although Jim's condition was caused by work, the company refused to accept his claim for workers' compensation, stubbornly insisting that the condition was not work-related and therefore not compensable under workers' compensation. When Jim could no longer work, he found himself out in the cold. He applied for unemployment insurance, but was told that he had to be available to go to work in order to receive benefits. He was too sick to work. The county welfare department finally agreed to give Jim and his family benefits, but urged him to apply for Social Security disability.

Jim and his family now eke out an existence on welfare because he is totally incapacitated. The family's total income is $350 per month and $180 per month in food stamps. Jim, his wife and two youngest sons, ages ten and twelve, had to move out of their modest but pleasant two-bedroom apartment in the Chicago suburbs to a welfare hotel in the seamier reaches of Chicago's North Side. The boys sleep in the living room. Jim hates the neighborhood and is afraid for his two boys. "Here it's a rathole. The kids don't learn too much at school. There's fighting on the streets, there's shootings, robbings, molestings, everything, you name it, window breaking, fires." As he talks fire and police sirens wail below, as if to illustrate his point.

The poverty of the neighborhood seems concentrated in the twelve-story building. It is the last refuge of those who, like the Richardsons, can no longer take care of themselves, a place where a family can stay together and survive, hoping for better times. The poverty is palpable. There are many small children in the lobby of the hotel and on the sidewalk in front of it. The children seem bright-eyed and better dressed than their parents. They don't yet know they're poor.

One four-year-old girl rides up and down in the ramshackle elevator. The old elevator man pretends that she's his assistant. The elevator is not summoned by pushing a button. There is a button on every floor, but over each one is taped a handwritten sign warning people not to push it. To call the elevator, people simply shout down the elevator shaft and call the elevator man by name. George shouts back to acknowledge the requests. At some floors the mechanism that releases the outside door to the elevator no longer works. The elevator man reaches over with a pair of pliers and releases the door. It takes a long time to reach the eighth floor, where Jim and his family live. The elevator becomes packed with people. Most seem to know each other and there is a lot of good-natured laughing and joking. All the people in the hotel are white, and I wonder if there is an exact replica of this building on the next block, where all the people are black. I'm sure there is.

The Jim Richardson who lives in this building grew up on a farm in Wisconsin, near Green Bay. He was always a healthy, strapping fellow. Before working at Schwinn his body was perfect except for a leg that he injured while parachuting when he was in the army, during World War II. Now he is able to sleep about four hours a night. And that he must do propped up at a forty-five-degree angle. He can't lie flat. If he does, he begins to lose his breath. He says, "I feel that I'm smothering, I can't get no air, I'm choking. If I'm not breathing right, sometimes I'll wake up and not be able to go back to sleep." Jim rarely goes out of the hotel. He can't walk for more than a block or two and he's afraid of being mugged. It would be worth his life to climb a flight of stairs. There is not very much the doctors can do to help him.

Jim has been shattered by his exposure to toxic chemicals at work. His body is a shell of its former self, and his life and that of his family have been thrown into the chaos of welfare poverty. A big, muscular man, Jim can barely sit for a whole hour. At the end of my interview with him he leaned over in his chair to tie his shoe. It took his breath away to perform this simple action. When he completed it he had to sit for a while. He wheezed and made little choking sounds until he was able to move again.

There is a striking similarity between the stories of Betty Smith, John Vandyne, and Jim Richardson. All three were exposed to toxic substances over a period of years: Smith to cotton dust, Vandyne to asbestos, Richardson to a variety of chemicals known to be harmful. These three workers were allowed to continue their exposure until an occupational disease had established itself in their lungs and nervous systems independently of any further exposure, until the damage was irreversible.

No one told Smith, Vandyne, or Richardson while they were still working that they were being slowly poisoned by their jobs. Their unions didn't tell them, their employers didn't tell them, and, as we shall see, their doctors didn't tell them. Oddest of all, none of the other workers told them of the danger.

But then, none of the workers seemed to know of the dangers of cotton dust and asbestos. In Erwin almost everyone has a father or grandmother or uncle or brother who died or was disabled due to a bad lung condition after working in the mill for many years. The

Erwin mill has been making cotton cloth since 1900. Why didn't the workers put two and two together?

The prosperity of Erwin, North Carolina, and its residents depends totally on the mill. Mary Lou Fisher, a VISTA worker who staffs the Erwin Brown Lung Association office, told me that the mill workers are still not interested in brown lung information. They won't listen to it and they still deny that cotton dust is harmful. When they retire or become disabled, these same people find their way quickly to the organization's office.

It appears that the psychological mechanism called denial is at work in Erwin. This same type of massive denial was found among Vandyne's fellow asbestos workers. It's not so strange that workers should want to deny to themselves that their jobs are dangerous. The worker does not create his own working conditions—they are given. The choice is to either put up with them or get out. If getting out means no job or a job paying a lot less, most workers will have little choice but to put up with industrial hazards. If the hazards are almost invisible, the exposure painless, and the damage takes years to be manifest—why, then, it is that much easier to deny their existence.

Richardson's case is a little different. In the Schwinn plant the toxic substances could be smelled and tasted. The danger was not silent and long term, as it was in the cotton mill and the asbestos plant. Yet the denial mechanism was still operative. Richardson, like many other workers, worked until he was physically unable to continue. He needed the job.

The medical profession has done little to dispel this denial. Very few doctors have the knowledge or motivation to properly diagnose occupational disease. Doctors tend to minimize what they don't understand.

Smith had symptoms years before becoming totally disabled. She was told by her family doctor that she had "bronchitis." It may well be that a family doctor in the denim capital of the world ought to be able to spot a case of byssinosis. This one apparently couldn't. When Smith came back to him with her diagnosis from a doctor in Durham, the family doctor told her that he himself didn't know anything about brown lung.

In not being able to diagnose Smith's condition as an occupational disease, Smith's doctor was typical of general practitioners across the country. Most of them are not familiar with even the most widespread occupational diseases. They rarely take a detailed occupational history from a patient. Their training has not prepared them to deal with this major health problem. American medical schools practically ignore occupational medicine. The average medical student gets one hour of instruction in this field in his entire four years in medical school.

Family doctors in urban areas are even more unlikely to diagnose an occupational disease than a doctor in a one-industry town like Erwin. It's unlikely that a city doctor will examine more than one or two patients who display similar symptoms due to similar exposure at work. Also, the exposure might have been twenty years before and 2000 miles away.

Large factories or mills always have in-house doctors or nurses. One would think that these medical professionals would be able to identify a disease characteristic of the plant's operation; they must run across it fairly often.

At the Burlington mill there was only an unsympathetic nurse. Her job seemed to be to dispense Band-Aids and call the rescue squad when a worker collapsed. Smith, like the other people there, took her medical complaints to her family doctor. In Vandyne's case there were regular checkups by Johns Manville doctors, including annual chest X rays.

These doctors saw the growing symptoms of asbestosis, especially the unmistakable X-ray changes. Yet, they did not warn Vandyne: they allowed him to work with asbestos until he became so disabled that he could work no more.

When doctors fail to diagnose an occupational disease there are legal as well as medical consequences. Not only does the worker not get proper treatment, if indeed there is any effective treatment, but he or she will not receive workers' compensation. In order to qualify for compensation, a worker must have a doctor's diagnosis of an occupational disease, a diagnosis that satisfies the workers' compensation requirement that an illness be work-related. Betty Smith could never have received even the small amount of compensation she did get while the diagnosis was "bronchitis." Luckily for Smith, organizers of the Brown Lung Association referred her to a pulmonary specialist in the university town of Durham. This doctor finally gave her the proper diagnosis, ten years after her first symptoms.

Doctors are a major bottleneck in the workers' compensation system. When the law was changed in North Carolina and byssinosis became compensable, there was not the flood of claims that one might have expected. Only a few hundred workers got compensation for byssinosis between 1963 and 1979. There were only a few doctors in the state who could diagnose the disease and feel competent to state unequivocally that a worker's disability was caused by exposure to cotton dust at work.

John Vandyne's case was a little different. The company doctors who examined him knew that he was developing asbestosis but did not warn him away from the work which they saw was harming him. Johns Manville's behavior in suppressing these findings was so egregious that the asbestos company's California employees were allowed by the California Supreme Court to sue their employer despite the heretofore insurmountable bar of the exclusive-remedy provision of the workers' compensation laws.

Richardson's experience was unusual among workers suffering from occupational disease. He found his way to a clinic that had been specially created to diagnose occupational disease and treat its victims. There are only five or six such clinics in the country, but they have had a salutary influence on other doctors in their localities, sensitizing them to the health hazards found in the workplace. These clinics provide appropriate treatment to workers made sick on the job and give them some chance to obtain workers' compensation.

Many occupations carry with them potential dangers as serious as those posed by cotton dust or asbestos. "No one died," the utility industry boasts, as a result of the accident at Three Mile Island. The statement is a little more than a rhetorical trick. The real issue involves the question of how many workers and nearby residents *will* die of cancer in ten or twenty years as a result of that accident. In luckier nuclear power plants, how many workers will die of cancer as a result of routine levels of exposure? How many chemical workers or farmworkers exposed to pesticides will develop tumors, or go blind, or wake up one day to find their limbs trembling uncontrollably? No one knows the answers to these questions, but when we look at what happened to people who worked around asbestos, we have good reason to fear the worst.

Occupational disease is fast becoming one of the nation's top health problems. There are at least two reasons for this. The first is that new technology is filling the American workplace with more untested and dangerous substances than ever before. Besides radiation, whose dangers are well documented, there are thousands of new chemicals being invented every year. Many of these are highly toxic. The other reason occupational disease is taken so seriously by public health experts is that medical science is beginning to identify as industrially caused much disease which was up to now thought to be nonindustrial. Many types of cancer, lung disease, nervous system disorders, and various other ailments are now suspected of being caused by conditions at work.

Public health problems, like all of society's troubles, cannot be solved until they are first recognized as "problems." While this recognition does not ensure that conditions will in fact get better, it is an absolutely necessary first step in the progress toward reform.

Only when the American public began to believe that there was an "environmental problem," that our air and water were becoming so polluted as to be a menace to our health and that of our children, did political and social change begin to occur. And changes did occur. The San Francisco Bay, which stank in the early 1960s, is now clean enough to swim in. Lake Erie has fish again. The air quality of American cities stopped deteriorating and even became better in many places.

Working conditions *can* be greatly improved for millions of people. Not made risk-free, of course, but as healthy as modern technology and human ingenuity can make them. Profound changes will occur when the issue of workplace health and safety becomes a major political issue, one that draws national media attention and delivers large numbers of votes to politicians bright enough to champion the cause.

There are many other stories of occupational illness among the 95% of those afflicted with it who do not receive workers' compensation; enough stories to fill many books: an encyclopedia of suffering.

There are cases which are truly bizarre. In Dawes Laboratories in

Chicago Heights, Illinois, male workers were exposed to large amounts of DES, a female hormone once given to pregnant woman. Human use of DES was discontinued because it was found to cause cancer in the female children of the users and genital abnormalities in male children. It was being manufactured as a cattle-feed additive. The male workers began to develop breasts and become impotent. One man's breasts grew so large that male hormone shots couldn't reverse the process. He had to have a mastectomy.

In upstate New York between 1975 and 1977, Eli Lilly and Company manufactured a pesticide that has since been taken off the market. Officials of the Oil, Chemical, and Atomic Workers Union, which represents the workers there, discovered an alarming fact. Between 1975 and 1979 there had not been one normal baby born to either a male or female worker there. Most of the birth defects involved the heart, and most of the babies died soon after birth. One baby has lived, but has had to have five operations to stay alive.

One of the most dangerous substances created by man is dioxin. If it is popularly known, it is because it was the most toxic ingredient in Agent Orange, the herbicide used to defoliate large areas of Vietnam. Also, there was a great deal of media attention to an explosion in a dioxin-manufacturing plant in Seveso, Italy, in 1977, which contaminated a large area. The effect of dioxin represents a summary of the more serious signs of a number of occupational diseases, particularly in the chemical and atomic industries: chloracne (a skin rash that looks like a bad case of acne and can last for years); liver damage; nervous system and vision impairment; spontaneous abortion; birth defects; abnormal function and derangement of the immune system; and cancer-causing potential—a catalogue of nightmares that is by no means exhaustive.

The frighteningly futuristic world of occupational disease coexists with the old-fashioned system of workers' compensation. The problems of the twenty-first century are being addressed with solutions from the nineteenth. It's no wonder that few occupational disease cases are compensated and that workers' compensation does nothing to encourage employers to maintain healthy workplaces. The type and degree of health hazards on the job that are now commonplace were not dreamt of when workers' compensation was introduced in the United States.

Workers' compensation changed the legal relationship between employer and employee. Before it was introduced, workers had the right to sue their bosses for negligence. Although such suits were difficult to maintain because of special defenses available to employers, at least the employee was in the same forum as everyone else who had a grievance against another. Because workers' compensation became the exclusive remedy, abrogating a worker's right to sue his or her employer, workers were expelled from the common law system and thrown to the tender mercies of state-run administrative agencies.

The law of negligence has evolved greatly since the early part of the twentieth century. The defenses that once defeated negligence suits are now just memories. The common law has shown remarkable adaptability to new technologies and new social forces. But these progressive developments don't help the worker who has been injured or made sick on the job. The worker is stuck like a fly in amber in the anachronism of workers' compensation.

4

A Gift From the Throne

Since the Industrial Revolution of the nineteenth century, Western society has been plagued by the problem of great numbers of workers hurt and disabled on the job. Workers' compensation was devised as the principal mechanism to solve or at least appear to solve this problem. Its purpose was to help workers survive the economic effects of on-the-job accidents and at the same time limit the financial liability of employers for those accidents.

Workers' compensation began as a simple, streamlined concept. Employees were to be compensated on a no-fault basis for on-the-job injuries. In exchange, they would relinquish their right to sue their employers for negligence.

This simple concept has grown over the past seventy years into an intricate complex of laws and institutions. Workers' compensation has become a maze of rules and legal interpretations filled with traps and dead ends for the unwary worker, windfalls for the doctors and lawyers who feed off the system, and an inexhaustible river of profits for the insurance companies that collect the premiums and pay the claims.

Workers' compensation is the only justification for immunizing employers from lawsuits of their employees. Workers have sold their birthright as citizens, their right to bring into a court of law those who have caused them injury, for a mess of pottage, the thin gruel of workers' compensation. How did it happen?

Workers' compensation is the granddaddy of social benefit schemes in the United States. It was adopted before any state had old-age insurance and its future was assured twenty years before Social Security became law. When workers' compensation was in its legislative heyday, between 1910 and 1921, unemployment compensation was an impractical dream, and regulation of child labor was often denounced as part of a socialist plot to undermine Americans' freedom of contract.

Although there was insurmountable political and legal opposition to other reforms at the time, workers' compensation achieved almost universal acceptance practically overnight.

Between 1911 and 1915, workers' compensation replaced the law of employers' liability in thirty-one states and the territories of Alaska and Hawaii. By 1921, forty-two states had workers' compensation laws.

At the turn of the century, workers' compensation was virtually unknown in the United States. Only a few academics interested in labor questions were aware of its existence and growing acceptance in Europe. Yet workers' compensation quickly gained almost universal acceptance.

The first major victory was won when it became the law of Imperial Germany in 1884. In 1897 it was adopted in Great Britain by the Tory ministers of the Kaiser's grandmother, Queen Victoria, and from there it made its voyage across the Atlantic, where it was made welcome by the managers of U.S. Steel and the National Association of Manufacturers.

Bismarck was responsible for the introduction of workers' compensation in the Second Reich. Not known as the friend of the German working person, or indeed of anyone besides his King/Emperor and fellow *Junkers,* Bismarck nevertheless created a system of social insurance that seemed to British and American observers as the last word in enlightened modernity. Workers were to be compensated for their work-related injuries regardless of fault, and their medical expenses were to be paid for them. In an era when the disabled worker in London or New York could look forward to begging in the streets, it is no wonder that the German system seemed like a most progressive model.

German "workman's compensation" was an integral part of Bismarck's antisocialist program. First trade unions were outlawed in the Reich, then in 1880 the Social Democratic Party. However, Bismarck was clever enough to realize that simple repression could not hold the lid on social and political change forever. But if there had to be change, let it come as a gift from the throne.

Having used the stick on their leaders, Bismarck offered to the German workers the carrot of workers' compensation (and other social insurance programs like old age insurance and unemployment insurance). He bragged publicly that his social policy was a way of bribing the working class away from socialism. As Kaiser Wilhelm I graciously noted in his imperial message to the Reichstag on November 13, 1881, "The cure of social ills must be sought not exclusively in the repression of social democratic excesses but simultaneously in the positive advancement of the welfare of the working classes." His Majesty hoped that the social reform laws would be seen by the Reichstag in their proper perspective as a complement to the laws banning socialism. Thus was born workers' compensation.

The first workers' compensation act in the English-speaking world was passed by the British Parliament in 1897 under the leadership of Joseph Chamberlain's Tory government. This legislation was based consciously on the German model; the Tories did not hesitate to credit Prince Bismarck for his ideas. Unlike Germany, workers' compensation in Britain was no gift from the throne. It was the product of a sharp political struggle that marked one of the more significant turning points in Anglo-American legal history. The result of the struggle was a new legal relationship between employers and employees. This relationship marked a sharp departure from the negligence principles that governed the law of personal injuries between almost all other parties. Employees and employees alone could collect from their employers for injuries on the job without regard to fault.

The political struggle that resulted in the birth of workers' compensation in England is worth considering. The two leading statesmen in British politics of the 1890s were Herbert Henry Asquith, leader of the Liberals, and Joseph Chamberlain, the mercurial Tory chief. When Asquith was prime minister he tried to pass an employers' liability bill that would have removed some of the major legal obstacles that prevented workers from suing their employers for on-the-job injuries. The bill passed the House of Commons in 1893 but was defeated in the more conservative House of Lords. The employers' liability bill was defeated because Asquith insisted on outlawing the practice of "contracting out." Contracting out was an agreement made between employer and employee to not be bound by the provisions of the employers' liability bill. In practice this meant that the employee could not sue his employer for any injuries he sustained on the job.

Many employers had made it a condition of employment to contract out of even the paltry rights of the 1880 employers' liability bill, the predecessor of Asquith's new, more forceful bill. No waiver of rights, no job. According to laissez-faire economic theory, employees would agree to contract out only in exchange for higher wages; therefore, contracting out was a reasonable bargain. The truth, obvious to anyone who had eyes to see, was that jobs were scarce and laborers plenty. A man or woman took what he or she could get. There was no equal bargain, just the usual case of the rich and powerful taking advantage of the poor and weak. The opponents of Asquith's bill representing business interests were willing to compromise on almost anything except contracting out. Asquith would not compromise on this essential principle, and the bill failed. *

The opposition won the election several years later, and the new Tory government passed the first workers' compensation act, a no-fault scheme. The Liberal Party, in eclipse for other reasons, did not oppose it. Organized labor in Great Britain had favored Asquith's employers' liability bill and opposed workers' compensation. Labor's greatest concern was to reduce workplace accidents, to make the workplace safer and healthier. The Industrial Revolution had left too many broken bodies in the wake of its inexorable march across England. The trade unions knew that if employers could be held liable in a court of law for the injuries of their employees, they would have a strong incentive to maintain a safe workplace. The trade unions did not favor workers' compensation because it was an

* David G. Hanes. *The First British Workmen's Compensation* Act. New Haven: Yale University Press, 1968.

insurance scheme that insulated the employer from the economic consequences of his employees' injuries. The trade union leaders were prescient. Workers' compensation has provided no real economic incentive for employers to clean up their workplaces. But British labor in the days of Joseph Chamberlain had not developed the political clout it was to have later. The workers' compensation bill of 1897 passed easily.

While the early English workers' compensation laws covered relatively few workers, they did establish the no-fault principle—and they paved the way for the elimination of the common law liability of employers for the injuries of their employees. "Contracting out" of liability was no longer left to the initiative of the employer–it was enshrined in the law. Workers' compensation can be viewed as one giant "contracting out" scheme. Therefore, Asquith's loss in the battle for the employers' liability act in 1893 is a decisive turning point in the history of workers' rights. The history is more relevant to present conditions in the United States than in Great Britain, since the country reformed its workers' compensation laws drastically after World War II.

Workers' compensation found fertile ground in the United States for the same reason it was accepted in Great Britain: there were so many killed and maimed on the job. The period 1903 to 1907, a time of prosperity, saw the highest industrial-accident rate in American history. The railroads alone were killing 3500 workers a year.

Also, the legal status of the American worker before workers' compensation was perceived as similar in all the states to that of the English worker. The worker could sue his employer, but the employer had defenses that made it difficult for the worker to prevail. Something had to be done.

The employers' defenses were created not by parliament or by state legislatures, but by judges. The common law is an intricate web of appeals court decisions that are published in official reports. A judge faced with a difficult case will find an earlier decision in a similar factual situation. No doubt feeling immense relief that he doesn't have to make up a new rule of law, he will follow the previous rule, justifying it on the legal concept of *stare decisis*, "let the decision stand." Of course, it is the legislature, or on the federal level the Congress, which makes the laws. The legislature can change the common law or abolish it altogether. The common law of worker accidents before workers' compensation appeared was abominable in England and not very good in the United States. Asquith's employers' liability bill would have altered the common law significantly. In the U.S. various state employers' liability bills were beginning to bring needed change. Workers' compensation swept away the common law altogether, thereby throwing the baby of personal-injury damages out with the bathwater of employers' defenses.

Three defenses, discussed below, formed the employers' bulwark against damage suits of their employees: the "fellow-servant doctrine"; "assumption of the risk"; and "contributory negligence." Of these, only the fellow-servant doctrine was restricted to suits of employees against employers. The other two applied to all negligence suits.

The three defenses were used by judges to deny workers damages even when there was no disputing the negligence of the employer. There was a good deal of agitation in the popular press in the United States in the years before workers' compensation became almost universal. The agitation focused on the injustice worked by the employers' defenses. Muckraking articles with titles like "The Cruelty of Our Courts" convinced the public that reform had to come. Teddy Roosevelt wrote a scathing article denouncing the legal doctrines that cast workers, destroyed by their jobs, into the gutter to beg or starve.

The Rough Rider's article was entitled "Sarah Knisely's Arm." It detailed the history of a lawsuit of one Sarah Knisely against her employer. One day in the early 1890s, Knisely's hand was caught in the cogwheels of a machine at which she was working. The machine must have been particularly dangerous because state law, even in those laissez-faire days, required the employer to put a guard on it. Knisely had complained to her foreman about the lack of a guard, but nothing had been done. Predictably, one day—or perhaps it happened at night, perhaps after she had worked twelve hours that day—the woman's hand was caught and her whole arm was mangled. She sued her employer. The court decided in 1895 that since Knisely knew about the dangerous condition of the machine, she had "assumed the risk" of its operation. This is the defense of "assumption of the risk." The through-the-looking-glass reasoning behind this decision goes like this: (1) The employer was negligent in maintaining an unsafe machine; (2) Knisely told her foreman that the machine was unsafe; (3) Therefore she admitted that she knew the machine was unsafe; (4) Nevertheless, she continued to work at the machine; (5) Therefore it was Knisely who was negligent. Knowing of the danger, Knisely should have refused to work near the machine. The law found that she had taken on the responsibility for the risk. She had "assumed the risk."

Never mind that there were others who would take her place if Knisely refused to work at the dangerous machine, that she would surely have been sent to work at a task just as dangerous or worse, or that she might have been fired. Never mind the fact that Knisely, like all factory workers of the time, lived hand to mouth and could not afford even a day off work, let alone enough time to try to find another job. When the case was reviewed by the Court of Appeals, the highest court of New York State, the honorable justices could not be expected to take these mundane facts into consideration. They did not. The year was 1895, and Sarah Knisley got nothing.

This was two years after Asquith's attempt to reform employers' liability had been defeated in Great Britain and two years before the British workers' compensation act was passed. It was fifteen years before the first New York State workers' compensation bill was enacted.

The "fellow-servant" doctrine was another defense available to the employer. If a fellow employee had negligently pushed Sarah Knisely into a machine that Knisely did not know was dangerous, she could not expect to get a penny from her employer. He could not be held responsible to her for any act of her "fellow servant." This pinched view of the employer's responsibility to his workers contrasted strikingly with his responsibility toward everyone else in the world. For as long ago as legal historians can trace, the doctrine of *respondiat superior* has been the law in English-speaking countries. This means simply that if an employee in the course of his employment injures someone, his employer is held accountable. In the 1830s and 1840s, judges in England and the United States carved out one major exception to this rule. They changed the common law so that the employer was to be held accountable to everyone in the world *except his own employees*. The fellow-servant doctrine was justly reviled by some of the progressive-era commentators who thought that workers' compensation was a way of eliminating it once and for all. It was not eliminated but, like an adaptable bacterium, has evolved into a basic principle of the system that was designed to replace it. Under workers' compensation, if an employee is negligently injured by a "fellow servant," he still cannot sue his employer, although he may claim workers' compensation benefits.

The last of the trio of defenses to a workers' lawsuit was the doctrine of "contributory negligence." If Sarah Knisely had not known that the machine was dangerous and had not been pushed into it by a fellow employee, she still might not collect a penny from her employer if she herself caused the accident through the slightest degree of negligence. No matter if her employer was 99% the cause of the accident and poor Sarah was 1%. In this case, the early-twentieth-century courts would proclaim that the victim's 1% guilt meant no recovery as a matter of law. Nothing for a mangled arm, an amputated foot, nothing for the family of the worker killed on the job if the employer could show that the worker was 1% or more responsible for the accident.

The terrible trio of employers' defenses to negligence suits by their employees were well known to the educated American public of 1910, or at least well known to those who read the newly popular muckraking magazines. Dramatic industrial accidents in those days were relatively common and their details were regular reading in millions of homes. One of the most appalling stories was the fire at the Triangle Shirtwaist Factory in New York in 1911. Over 140 women who worked in the factory died, primarily because the owners of the company had kept the fire doors locked and bolted so that the workers couldn't take breaks on the fire escapes. Train wrecks were not uncommon, and the accidental death of railroad workers occurred with monotonous regularity. In 1913, 25,000 workers in America were killed on the job and another 700,000 were seriously injured.

Workers who were injured on the job or the families of dead workers could sue their employers if there was any negligence on the employer's part. There usually was, but the worker or the worker's family would probably have to wait for at least a few months before getting any money. In 1910 there were not the public benefits available which today differentiate our society from the Third World, where it is possible for a family to sink to a level of poverty so deep that small children must be sent out into the streets to beg or the family will starve to death. Small children did beg in New York in 1910, and some of them had fathers and mothers who could not work because of industrial accidents.

There was no Social Security, no food stamps, and no Aid to Dependent Children. Few workers had savings to fall back on.

The workers' compensation decade in America, from 1910 to approximately 1920, was the high-water mark of the progressive movement. It was a time of reform, of experimentation. Reform was certainly needed. The great corporate and banking tycoons appeared to dominate state legislatures, the courts, Congress, and even the President. The dream of Jeffersonian democracy, the small independent farmer, was finally laid to rest. Between 1900 and 1920 the urban population in the United States grew 80%, while the proportion of the population engaged in farming declined correspondingly. A large part of the urban population worked for the few great capitalist enterprises. The great army of industrial wage earners lived hand to mouth, earning subsistence wages and working up to ninety hours per week. This burgeoning city population seemed to many middle-class, native-born Americans to represent a strange and disturbingly new way of life. The progressive movement was an expression of middle-class revulsion at the abuses of the raw capitalism of the trusts.

Progressivism appealed not to what we now call blue-collar workers but to small businessmen whose economic existence was threatened by the trusts, to professional people who were appalled by the crassness of the industrial magnates and the oppression of the workers, and to many farmers who were dependent on railroads owned by recklessly greedy monopolists.

The progressive movement was really a state of mind. It could be defined as much by what it opposed as by what it favored. It was not coincidental that progressivism's opponents were called "standpatters." Progressives welcomed change and new ideas. In 1911, a representative of the National Association of Manufacturers, himself a self-styled progressive, would condemn the reactionaries in his own organization. Reactionaries resisted change on principle. They wanted only to protect what they had. Progressives were not afraid of government involvement to bring about reforms that they felt were important. The stand-patters, the reactionaries, believed in the pure laissez-faire doctrines of the Manchester School. To stand-patters, the role of government was exceedingly limited. It was they who denounced the idea of state-imposed limitations on the number of hours children could work as a sinister socialist plot, radical and un-American. The progressives favored a list of social reforms that included workers' compensation. This modern foreign idea appealed to reformers who believed in innovation and positive state involvement in bettering the lot of the laboring classes. Workers' compensation on the English model would replace the complicated judge-ridden system of employers' liability, with a brand-new, rationalized, simplified system of compensation. As in the Germany of Bismarck and the England of Joseph Chamberlain's time, neither the mass of workers nor the labor unions had much to do with the introduction of workers' compensation. The progressives did not have any significant following in the working class, and for the most part were not particularly friendly to organized labor. They thought they knew what was best for those on the lower rungs of America's social ladder.

Progressives felt that something could and should be done about the horrific toll of industrial accidents. The injury and death rate had to be reduced. There was no disagreement on that point. Injured workers and the families of dead workers should be compensated in some way. Surprisingly, there was little disagreement on this point. American workers were being crushed under a hodgepodge of archaic laws. Or were they? What in fact was the legal situation of American workers who had been injured on the job in the years 1910 to 1915, the beginning of the decade of virtual explosion of workers' compensation laws?

The law was changing. Sarah Knisely's cause was beginning to be vindicated. In 1912, New York State's highest court, the Court of Appeals, overturned its own 1895 decision and severely restricted the defense of assumption of risk. Teddy Roosevelt called this decision a victory of an aroused public opinion. Judges read newspapers, too.

By 1907, twenty-six states had passed employers' liability laws that ameliorated the harshness of the fellow-servant rule. Some of these laws also eliminated assumption of the risk and contributory negligence. The courts, following public opinion, were finally opening up to injured workers, and injured workers and the families of workers killed on the job were actively beginning to win negligence cases against employers in significant numbers.

If workers had an uphill battle against a hostile legal system, they had one great advantage over their employers when they got into the courtroom. Juries in the progressive era tended to be sympathetic to injured workers. At a time when big business was as popularly disliked and distrusted as it was powerful, the worker who was in court to sue U.S. Steel or Standard Oil for his injuries could count on getting a sympathetic ear from a jury of his peers. Jurors were more likely than judges to come from the same social class as the worker/plaintiff. Jurors tended to understand the economic bind that forced a worker to do a dangerous job, or the overpowering fatigue at the end of a fourteen-hour day that led to a fatal misstep. Many unsophisticated workers and their families accepted tiny settlements because they had no conception of their rights. However, the ones who did go to court in the period 1910–15 had good prospects.

Damage awards for injured employees just before introduction of workers' compensation were relatively generous. A study by Richard Posner, a professor of law at the University of Chicago, showed an average award where an employee lost a limb or had an equivalent injury in 1905 of $10,138. * When we consider that railroad workers earned at that time about $500 per year and textile mill workers about $300 per year, this average award appears reasonable indeed. Workers' compensation put an end to these relatively large awards once and for all.

American workers might have had the sense that a new day was dawning when Congress passed the Federal Employers' Liability Act of 1908. The act was a great advance for workers. It abolished the fellow-servant doctrine and, in most cases, assumption of risk. It eliminated contributory negligence as an absolute defense and adopted the principle of comparative negligence, so that the worker's own lack of care reduced the amount of damages he could get, but did not let the employer get off scot free. The Supreme

* A Theory of Negligence." *Journal of Legal Studies*, volume 1, January 1972, p. 29.

Court held this law constitutional in 1912, and a flood of lawsuits ensued. It is still the law governing interstate railroad employees, who have fought off workers' compensation since then.

The legal logjam that had blocked workers' access to the courts was beginning to break up in the period just before the introduction of workers' compensation. New state employers' liability laws, favorable jury verdicts, the Federal Employers' Liability Act, all backed by a vociferous and awakened public opinion, indicated that the old legal relationship between "master and servant" was undergoing a dramatic shift. The heavy economic burden of industrial injury was beginning to shift from the backs of the workers to their employers. This development was stopped dead in its tracks by the introduction of workers' compensation, the gift of Kaiser Wilhelm to the world.

Workers' compensation swept away the common law of employers' liability and the employers' liability statutes in an astonishingly short time. In 1910 workers' compensation was a visionary scheme of progressive reformers. In 1920 it was entrenched in the laws of forty-one states. No other reform of the progressives achieved a success as rapid and as far-reaching. Workers' compensation triumphed in the United States because of a temporary alliance of progressives and conservatives, a congruence of interest of small business and big business. It triumphed at a time when organized labor was fighting for its very existence.

A political or social idea can be best understood by looking at its friends and champions. Two of workers' compensation's staunchest supporters in 1912 were U.S. Steel and the National Association of Manufacturers. U.S. Steel was formed in the early years of the century. It was created out of the major competitors of the American steel industry. U.S. Steel was the trust to end all trusts in the American steel industry. Although small steel companies continued to exist, U.S. Steel had virtually no competition. It *was* Big Steel, and it dictated standards for the entire industry.

In 1909 and 1910 U.S. Steel made two significant decisions. First, it announced a hard-line antiunion position. The steel corporation simply would no longer even talk to the representatives of any union, despite the presence of a union of skilled workers in the steel industry dating back to the 1880s. Coupled with this announcement was a positive effort to destroy all types of union organizing activity. The corporation declared war on unions. U.S. Steel spies reported to management the names of workers who were trying to organize or who were sympathetic to union organizing efforts. These unfortunates were fired and blacklisted. This form of economic capital punishment ensured that they would never work in a steel mill again.

U.S. Steel's strategy was simple—its managers wanted to maximize profits by keeping labor costs at a minimum—and 1909 was a logical time to declare war on unions. The union of skilled workers that had been relatively influential since the 1880s was now seriously weakened. Great advances in technology had lessened the need for skilled workers. At the same time, a tidal wave of immigration filled the ranks of the unskilled to overflowing. Labor was cheap and plentiful. U.S. Steel could fill its factories with little trouble. The unions, however, could not or would not organize the armies of the unskilled. The corporation clearly had the whip hand and knew how to use it.

But the men running U.S. Steel in 1909 and 1910 were not mere Neanderthals, throwbacks to the reckless robber barons of the previous century. Many of them considered themselves progressives. They believed in rationality and were aware that they could not just take from the workers and give nothing.

In 1910 U.S. Steel announced a voluntary program of workers' compensation. Based on the no-fault principle, this program would provide some degree of economic security for workers injured on the job or the families of workers killed on the job. U.S. Steel announced that its program was instituted as part of the company's relief policy and not because the company had any legal obligation to its workers. Company spokesmen were anxious to point out that this program was not the result of any demand or suggestion by its employees.

Like Kaiser Wilhelm, U.S. Steel had recognized that it was not enough to repress the opposition. A modicum of reform was needed to prevent really significant changes.

U.S. Steel might have had another consideration. By taking advantage of the company's benefits, a worker would be found to have waived his rights to sue U.S. Steel for injuries as a result of the company's negligence. In a time of rising jury awards and public hostility to the trusts, U.S. Steel's workers' compensation program made good economic sense—at least to its own board of directors and stockholders.

U.S. Steel's antiunion, pro-workers' compensation policies were nicely complemented by the activities of the National Association of Manufacturers, which represented medium- and small-scale industrialists. By 1912, NAM had made progress in promoting workers' compensation. It had done even more to hurt organized labor. The NAM was a leading proponent of workers' compensation. In 1909 the organization authorized creation of a committee to study the problem of compensation for workplace accidents. The committee surveyed 25,000 American employers. One half of those questioned responded, and of these an astonishing 98% were in favor of "automatic" compensation for employees injured in their employment.

Members of the committee traveled to Europe to study workers' compensation in Great Britain and Germany. They returned enthusiastic about the system. This system of compensation appeared to be well accepted by "responsible" people in those countries, and there seemed no good reason why it could not be transplanted to the United States. The committee staff drafted a model workers'-compensation statute, which the National Manufacturers' Association used in its lobbying efforts.

Workers' compensation was of course not the only interest of the NAM in this era. Even dearer to the hearts of its members was union busting. In 1903 the NAM became the leading force in an all-out propaganda war against unions and the closed shop. NAM speakers scoured the country warning the good citizens against the "treason" of people such as Samuel Gompers, the highly respectable leader of the American Federation of Labor.

The NAM circulated free pamphlets warning of the dangers of the eight-hour workday, the right of every type of man to sell his labor in a free and unrestricted market, and the destructiveness of the union label. The drumbeat of antiunion sentiment probably reached its apotheosis in a speech given by John Kirby, the president of the NAM, to the students of Kenyon College in Ohio in 1911. Kirby accused the AFL and its leaders of everything from maiming and murdering "thousands and thousands of peaceable workers" (strikebreakers) to plain "bad citizenship." Gompers and other top moderate labor leaders he accused of treason. Kirby said that labor unions favoring the closed shop should not be allowed to exist and that the fundamental question of labor relations is "who shall operate business," those who own it or the employees? There will be no industrial peace in the United States, he said, until this fundamental question is decided. Reaching a crescendo of hysteria, Kirby declared that if need be the U.S. government should press into service the whole of its army and navy to preserve the rights of owners.

The organization of which John Kirby was president, the NAM, was one of the principal backers of workers' compensation.

After 1910, workers' compensation was viewed by contemporaries as a "progressive idea." It was an innovation, at least in the United States, and it required legislative involvement in solving a social problem. Furthermore, workers' compensation, with its rejection of the negligence principle, seemed streamlined. It resolutely eschewed the gewgaws and ruffles of the common law. It seemed, well, so twentieth-century.

Organized labor was by and large opposed to workers' compensation until about 1910. Sensing that the movement for it was inexorable, labor began to support the system while trying to enlarge the scope of employers' liability. Organized labor did not realize until it was too late that there would be no employers' liability, that workers' compensation would become the "exclusive remedy."

Businessmen, both big and not so big, came to see eye to eye with the progressives. Their notions were hardly altruistic. They knew that the frightening toll of on-the-job accident victims was beginning to fuel movements for reform. But if reform had to come, let it come as a gift from U.S. Steel and the NAM. As the chairman of NAM's workers' compensation committee put it in a speech to an insurance industry group in 1912, "Unless we . . . help to settle these problems . . . they will be settled for us with a vengeance by the agitator and the demagogue." The old employers' defenses in common law were crumbling. Big business was hated and could expect little sympathy from juries in negligence cases. Workers' compensation offered the prospect of an easily insurable risk and a guarantee that the employee's recovery would be strictly limited. The cooperation of big business and progressives, and the acquiescence of leaders of organized labor, ensured the rapid adoption of workers' compensation by state legislatures across the country.

5

Workers' Noncompensation

The original goals of workers' compensation appeared eminently reasonable. There is little doubt that many of the progressive reformers who advocated workers' compensation thought that the system they were creating would sweep aside the archaic laws that governed relations between "master and servant," at least as to accidents on the job, and bring order out of chaos. What the U.S. Steel executives thought workers' compensation would achieve can only be surmised.

Workers' rights to sue for negligence were given up because people believed that workers' compensation could achieve its ostensible goals. If workers' compensation has not achieved its ostensible goals, if in fact it has merely replaced one form of injustice with another, then there is no rationale for workers to be denied the same right to sue their employers as the consumer has to sue the manufacturer of a defective chain saw which ripped open his arm.

There are six basic goals of workers' compensation. The success of the system can be measured against these. Workers' compensation was supposed to:

1. *Internalize the costs* of industrial accidents in the industry that caused them. As Lord Asquith said, "The

cost of the product bears the blood of the workman." Internalizing costs encourages employers to maintain a safe and healthy workplace, since their margin of profit would depend in part on how much it cost to compensate their workers for accidents. Employers with safer workplaces and hence fewer compensable accidents would have a competitive edge over employers who were so negligent as to allow their workers to be hurt.

2. *Compensate workers without regard to fault* in causing the injury. The purpose of this was to eliminate litigation. Reformers felt that since the main issue in personal-injury litigation was the question of who caused the injury, to get rid of that issue was to eliminate the need for litigation.

3. *Provide medical care* for the workers' on-the-job injuries. Before workers' compensation, the injured worker was responsible for paying his or her own medical bills. Even if he finally got a judgment in a suit against his employer, or settled the case out of court, the money would usually come too late to obtain good treatment. In practice, injured workers usually had to rely on charity.

4. *Provide wage-loss replacement.* Workers injured on the job were to be given a percentage of their wages, from 50% to 66⅔%, while they were unable to work. Thus, they would not have to rely on charity.

5. *Provide swift and certain compensation.* Compensation for on-the-job injuries would no longer be subject to the vagaries of juries and the delays of a clogged court system.

6. *Limit the liability of employers.* Large personal-injury awards could ruin a business. By eliminating the employees' right to sue, workers' compensation reduced the employers' risk so that it was easily insurable.

Workers' compensation was enacted into law by all of the states in order to achieve these goals. Originally, it was little more than direct and very limited cash benefits. It was soon apparent that a more elaborate system of compensation was needed to take care of workers' needs and to prevent workers from getting compensation to which they were not entitled. The system or rather systems (each state has its own compensation laws and customs, although they bear a strong family resemblance to each other) have grown luxuriantly. In order to decide whether workers' compensation has met or come close to meeting its original goals, it is first necessary to have a basic understanding of how it functions. Although workers' compensation is closely structured by state laws, it is a system of private insurance, offering a number of different types of benefits.

Workers' compensation benefits offered fall into six categories. They are:

1. *Medical benefits*. In most states all medical bills for work-related injury or illness are supposed to be paid by workers' compensation.

2. *Temporary total-disability payments*. Designed to replace wages lost while an injured worker is recovering, these benefits vary greatly from state to state.

3. *Permanent total-disability payments*. This category of benefits is supposed to replace future income lost by a worker who is unable to work again because of industrial injury or occupational disease.

4. *Permanent partial-disability*. PPD should represent the difference between the income a worker could earn with a job-caused disability and his previous income.

5. *Death benefits*. These are paid to the family or dependents of a worker killed on the job.

6. *Vocational rehabilitation.* This category of benefits, required in only a few states, is designed to help a partially disabled worker find a new trade or career so that he or she can become once again a productive member of society.

These benefits are not a matter of contract between an employer and an insurance company. They are carefully spelled out in state law and may not be altered except by action of the state legislature.

Benefit levels and rules regarding eligibility vary greatly from state to state, but some important generalizations can be made about workers' compensation in almost all states. Benefit levels are too low. It is too difficult to establish eligibility in certain kinds of cases, particularly occupational-disease cases and back-injury cases, and adjudication procedures range from adequate to abysmal in terms of efficiency and fairness. In most states there is no attempt made to help a disabled worker return to the workforce.

Workers' compensation generates a great deal of litigation. Insurance companies dispute a high percentage of claims based on lost-time accidents. The percentage of claims disputed rises in relation to the seriousness of the accident or occupational disease. The more serious the accident or disease, the more chance that the claim will be denied and have to be litigated. The burden of litigating a case falls squarely on the injured worker.

In order to get any workers' compensation benefits, a worker must be able to prove that his or her injury is work-related, that it "arises out of and in the course of" employment. In California workers' compensation slang, the injury must be "AOE/COE." Although the wording of the legal requirement changes slightly from state to state, the concept is the same. The injury or illness, to be compensable—that is, to be covered by workers' compensation—must be in some way caused by a job. Furthermore, the injury or disease must be traceable to a particular job so that the proper employer and insurance carrier can be held responsible.

If the employer or carrier denies liability, the injured worker must request a hearing with the state agency empowered to decide workers' compensation cases. Until that hearing is held and the state official orders the employer or the insurance carrier to pay, the worker must simply wait. Obtaining a hearing may take from six weeks to a year, depending on the state. Massachusetts has only twelve officials empowered to hear all workers' compensation cases, Indiana only six. The difficulty of getting a hearing and the frustration for the injured worker can be imagined. This dismal state of workers' rights contrasts vividly with the constitutional right of applicants to be put on the welfare rolls immediately on application for benefits and to be kept on the rolls while determination of eligibility is made. Of course, the distinction between workers and welfare recipients is an artificial one. Many people on welfare are workers who are waiting for workers' compensation benefits or for whom workers' compensation has been denied. Many people on welfare in America were workers whose compensation has run out, and who cannot work because of job-related disability.

The requirement that injury and consequent disability be work-related is a reminder that workers' compensation still bears features distinctive of a litigation system. There are "plaintiffs" and "defendants" in workers' compensation cases and there are findings of responsibility to be adjudicated. It is not enough to be injured or disabled. The worker must prove that the disability was related to a particular job and must have correctly named as defendants the employer and its insurance carrier. This contrasts with government benefit programs such as SSDI (Social Security Disability Insurance Program) or AFDC (Aid to Families with Dependent Children), where litigation is incidental to the system and is usually limited to the issue of eligibility.

Doctors play a pivotal role in deciding whether an injury or disease arises out of and is in the course of employment (AOE/COE) and of a particular job. Medical reports customarily decide a worker's legal fate.

Many accidents are so obviously work-related that the AOE/COE concept is not subject to dispute by the employer or the insurance carrier. When Joan Thomas' arm was mangled while cleaning a conveyor belt in Iowa, there could be no doubt that the injury arose out of and in the course of employment. Other types of injuries, and particularly occupational diseases, are harder to establish as AOE/COE.

Back injuries can also be hard to establish as AOE/COE. Mike Sarkis was luckier than many. The insurance carrier covering his employer accepted his back condition as AOE/COE from his last job because of their own doctor's diagnosis. Often workers who have developed bad backs on the job do not happen on a doctor who recognizes the condition as work-related. Many doctors think that back conditions are due to a natural aging process, and they usually don't take thorough occupational histories. The result of a diagnosis that the back condition is not AOE/COE is denial of the workers' compensation claim.

The medical deck is stacked against the worker to begin with. In urban areas there have grown up over the years "industrial medicine" clinics. The doctors in these clinics are paid by employers and insurance companies and are often responsive to their customers' needs. These needs are as much legal as medical. On the one hand these doctors treat injured workers and on the other hand they provide the medical opinions that will serve the needs of their customers (not their patients) when the time comes for workers' compensation litigation. Bethlehem Steel sent Dorothy Hanna, the Indiana steelworker, to a doctor for a special examination. His first question to her was: "Do you plan to sue the company?" For once Dorothy was speechless.

There are, of course, doctors on the other side, ones who diagnose every condition as work-related and who get their patients as referrals from plaintiffs' lawyers. These are rarer. "Industrial medicine" is largely dominated by industry.

No more than 5% of those Americans disabled due to an occupational disease in 1972 received workers' compensation. There are two main reasons for this amazingly low percentage. The first is simply that many workers don't know that the illnesses from which they suffer are due to exposure at work and are therefore potentially compensable under workers' compensation. Most cases of what is actually occupational disease are never reported as such and claims for compensation are never filed for them. But even if a claim is filed, it has a very poor chance of resulting in benefits for the sick worker. Workers' compensation carriers contest a great percentage of occupational disease cases and, for a variety of reasons, usually prevail. Insurance companies win these cases because it is impossible in most instances for the worker to prove that the disease from which he or she is suffering arose out of and in the course of employment, that the condition is AOE/COE.

A disease or illness caused by work may be either a specific occupational disease or one that is found in the general population, but caused by work. An example of a specific occupational disease is byssinosis, the disease that struck Betty Smith and her fellow cotton mill workers. Cancer is a disease that affects the general population but that is often caused by exposure at work. Except for some very rare types of cancer such as mesothelioma, which is caused only by exposure to asbestos, cancer is particularly difficult to establish as work related since it may be caused by exposure to a multitude of substances, many of them found in the general environment.

Whether the disease is typical only of a certain occupation or is relatively common, chances are that it has a long latency period, a long time from exposure to a hazard to first appearance of symptoms of the illness. The latency period is particularly long for cancer. It has been estimated that the average latent period for the development of lung cancer due to asbestos exposure is eighteen years. There is also a great variation in the latency period for different people; some seem to develop diseases of occupation much faster than others.

Most people don't connect their newly developed disease with work done years before. It never occurs to them that they were exposed to a hazard five to twenty-years before and that their present condition is the direct result of that exposure. Workers usually don't know what they are exposed to, and even if they do they only rarely understand the causal relationship. These workers can expect no help from doctors in finding the real cause of their illnesses. Few doctors know anything about occupational diseases, and few doctors bother to take detailed occupational histories.

But there are the John Vandynes and the Betty Smiths and the Jim Richardsons who have found their way to doctors who have been able to diagnose their conditions as work-related. What are the hurdles they face in trying to establish that their conditions are AOE/COE?

The worker who suffers from an occupational illness has the burden of proving that a certain exposure at work caused a certain disease. He or she must be able to show that the exposure *probably* caused the disease—it is not enough to show that the exposure *possibly* could have caused it.

Smith in North Carolina, like Vandyne in California and Richardson in Illinois, had to prove a number of facts in order to establish that a specific exposure caused a specific condition. She first had to show that she was in fact exposed to a substance that is toxic. By the middle 1970s even the most diehard cotton mill owner had to admit that cotton dust was toxic. Vandyne's case is similar. By the time his case was filed, no one could seriously doubt that asbestos was toxic. Of course, both cotton dust and asbestos were known by scientists to be harmful many years earlier.

Richardson faces a much harder task than Smith or Vandyne. He doesn't even know what he was exposed to. His former employer won't tell his doctors what chemicals he worked with and there is no law requiring disclosure. But even if Richardson finds out, he still must be able to prove that the substance is toxic and that it is known to produce the kind of disease he suffers from. But there are thousands and thousands of chemicals used in the workplaces of America and only a few of them have been studied with the degree of scientific rigor that would stand up to the legal requirement that a particular exposure must be shown to have probably caused a particular disease. Furthermore, Richardson was exposed to many chemicals in a variety of combinations. No one, not even Richardson's employer, could know in what combinations.

After proving exposure to a toxic substance, the worker must prove that the dose he or she was exposed to was high enough to cause the disease complained of. This is usually impossible.

The worker must be able to show that it was the toxic substance to which he was exposed at work that caused the disease and not some other non-work-related exposure that caused it. Spokesmen for the asbestos industry blame tobacco as the cause of lung cancer among asbestos workers. The kernel of truth in this assertion is that workers who smoke have a much higher chance of developing lung cancer than those who do not. It is well known that the combination of smoking and exposure to asbestos is very hazardous. Nevertheless, studies have shown that smokers who work with asbestos have a much greater chance of contracting lung cancer than smokers who do not work with asbestos. In a particular workers' compensation case it is in practice up to the prejudices of the "judge" whether to believe that lung cancer was caused by asbestos exposure or by smoking.

With the exception of a few combinations like smoking and asbestos exposure, little is known by scientists about the effects of toxic substances when combined with each other or with other substances that are nontoxic by themselves. Experiments to establish toxicity almost always isolate a substance and test its qualities without regard for its effects when combined with other substances. Most workers have been exposed to a variety of substances, some toxic in themselves, some toxic only when combined with other substances. There are diseases caused by these combinations whose origins are impossible to prove. There just isn't enough scientific evidence to prove that substance X causes cancer only when combined with substance Y. It is up to the worker to prove the causal connection. The employer or the insurance carrier need do nothing and will win the case if the worker cannot bring in the evidence to prove the case.

Even if the worker suffering from a disease contracted on the job is able to find out exactly what he was exposed to and in what quantity, and show that credible scientific studies have proven that the exposure causes the disease he's suffering from, he still faces laws in many states that are meant to prevent him from being compensated. Thirty states exclude from workers' compensation coverage "ordinary diseases of life." Thus, Betty Smith, who suffers from a disease contracted only by those who work with cotton, might get compensation if her diagnosis is "byssinosis" but not if it is "bronchitis," which is found among the general population and thus considered an "ordinary disease of life." In those states occupationally caused cancer, except for a few rare forms, is not compensable. *

Statutes of limitation are also designed to defeat occupational-disease claims in many states. In New York, compensation is payable for occupational disease only if incapacity results within three years of the last exposure to the hazard that caused the disease. This type of limitation period is found in many states. Vandyne would be out of luck in one of them. He hadn't worked around asbestos for at least six years before the symptoms of asbestosis appeared. The average latent period for cancer resulting from exposure to asbestos is eighteen years. If, during this latent period, the exposed worker is

* For a good discussion of legal barriers to compensation of occupational disease, see Peter Barth, *Workers Compensation and Work Related Illnesses and Diseases*. Cambridge, MA: 1980, pp. 94–99.

no longer working around asbestos, he has no claim in states like New York, no matter how perfect his evidence that the exposure caused his illness.

The only surprise about the figure of under 5% of workers disabled due to occupational disease receiving workers' compensation in 1972 is that so many managed to jump all the nearly impossible hurdles put in their way by workers' compensation laws and procedures that seem meant to defeat occupational-disease claims.

That it is not impossible for a system to be devised that could compensate for occupational disease can be seen by a look at the achievements of several foreign countries, as shown in the chart below.

Table 1 shows the number of occupational-disease (OD) cases actually compensated in certain foreign jurisdictions and extrapolates from that the number of cases that would be compensated if the jurisdiction's working population were as large as that of the United States.

Table 1
Comparison of Workers' Compensation for Occupational Disease Among Six Foreign Countries and the U.S.

Jurisdiction	*Number of Disabling OD Cases Compensated*	*Number of Compensated Claims Weighted to Reflect Size of U.S. Workforce*
Sweden	16,000	340,000
Ontario, Canada	5032	140,000
Belgium	5737	140,000
Switzerland	4047	125,000
Great Britain	14,499	50,000
France	8080	40,000
U.S.A.	30,000	30,000

Source: ASPER, U.S. Department of Labor, 1979.

Suppose a worker gets a foot in the door and his or her injury or disease is accepted or adjudicated as arising out of and in the course of employment. Suppose that the "temporary-disability" period is over. The worker's condition is now what they call in California "permanent and stationary," meaning that the injury or condition is not going to change one way or the other. Suppose finally that there is a residual disability, not total disability (the worker in a wheelchair, for example,) but just enough disability to prevent the worker from doing a great deal of what he or she could do before, especially as far as work is concerned. A right-handed carpenter who has lost the use of his right hand, for example. How does workers' compensation make up for the loss?

Compensation for partial disability is called "permanent partial-disability." It is the only benefit offered by workers' compensation to make up for whatever permanent losses are suffered by a worker as the result of an industrial accident. The only exception to this is the category of benefits called "permanent total-disability." Permanent total is rare. In 1978 there were only 2600 awards for permanent total-disability in the whole country. In the same year there were 418,000 permanent *partial* awards. There were many more permanent *partial* cases settled between the parties without recourse to a formal hearing. There probably were not many more permanent-*total* settlements, however, since workers' compensation insurance carriers tend to contest the most serious cases with a vengeance.

The states have evolved different ways of calculating how much money a partially disabled worker is entitled to, but the most common method is a chart of losses, a workers' compensation "schedule" of benefits. The schedule expresses benefits in terms of weeks of disability payments. In Texas, the third-largest state in population, the loss of the right foot is worth 125 weeks of compensation. The maximum amount of weekly compensation is as of 1982 $140 per week. Thus, the maximum amount a worker could collect for the loss of his right foot in Texas would be $17,500. These dollar amounts apply whether the foot is actually cut off in an accident at work or merely rendered useless.

Table 2 shows the *maximum* amounts that may be awarded to a worker in Indiana for the loss of certain specific parts of his or her body.

Table 2
Indiana Permanent Partial-Disability Schedule

Loss of thumb	$4500
index finger	3000
little finger	1500
Hand Below Elbow (sic)	15,000
Foot Below the Knee (sic)	13,125
Loss of Both Testicles	11,250

Source: "Workers' Compensation (Indiana Edition)" United Automobile Workers Region 3.

Joan Thomas' arm, mangled in a conveyor belt in Iowa, was not worth much in the workers' compensation scheme of things. After her nine operations she was able to use it pretty well—she didn't lose much function. That the insurance carrier gave her almost nothing for permanent partial-disability, $2000, was not really shady. She might not have gotten a whole lot more if she had litigated the case.

There are many cases that do not lend themselves to schedules. Back cases, occupational-disease cases, and psychiatric cases force the workers' compensation system to consider the effect of an injury or disease on the worker as a whole person. Permanent partial-disability is determined in these cases by evaluation of medical reports that are written in code words that can be translated into percentage points of disability, which in turn are automatically translated into dollar amounts.

Mike Sarkis was examined by an orthopedist chosen by the insurance carrier and one chosen by his own lawyer. Not surprisingly, the insurance company doctor wrote that Mike's disability due to his back condition was not really too bad. He could no longer do "very heavy work." Mike's own orthopedist said that Mike was now limited to "light work" because of his job-related back condition. The insurance company's doctor's evaluation was worth a 15% rating in California, or $3517. Mike's doctor's rating was worth a 50% rating, or $16,870. Mike's lawyer sent him to another orthopedist agreed upon by both sides, an "agreed medical examiner." This worthy opined that Mike could no longer do "heavy work." "Heavy work" is a code phrase for a 30% disability rating, worth $8452.50. That's what Mike finally got.

The purpose of permanent partial-disability is to provide a disabled worker substantial protection against loss of income. Permanent partial-disability should make up the difference between what a worker earned before being hurt on the job and what he can earn now with his permanent disability. That's the theory, anyway.

Workers' compensation permanent partial-disability, with its schedules and its percentage ratings, has strayed far from the principle of income protection. The $8452.50 that Mike Sarkis got for his disability will not make up for even two years of lower wages, let alone a lifetime of lower wages. But that's just what Mike has to look forward to—a lifetime of lower wages.

Permanent partial-disability does focus the attention of the injured worker on the prospect of getting some kind of cash compensation for an on-the-job injury. Injured workers and the general public are lulled by PPD into thinking that workers' compensation is in fact a *compensation* system. As Congressman George Miller (D-Cal.), Chairman of the House Labor Standards Subcommittee, told me in an interview in 1981: "Workers' compensation keeps the workers from organizing. Each one's treated as an individual case so that each worker goes into a system that's inadequate and has his personal experience and by the time it's all resolved he's out, he's out of the union, he's inactive or he's out of a job, he's not seeing his friends anymore, and he's not going to organize. The belief of each worker is that in fact the system does work, until they engage the system. It's like inadequate payments of welfare. They sort of keep the people quiet. That's all that's going on. People aren't getting compensated for their loss."

Not only is permanent partial-disability not enough, it is also hard to get. This category of workers' compensation benefits is the largest single reason for litigation. The majority of cash benefits paid out by workers' compensation insurance carriers is for permanent partial-disability payments. From the insurance industry point of view, big money is involved, and big money is not given up without a fight. Less than 30% of workers' compensation cases involve payments of cash benefits; most are just medical payments. But of the almost 30% of cases involving cash payments, 60% of all the money paid out is for permanent partial-disability. This is an extraordinarily high percentage, considering that the other 40% of cash benefits includes the combined total of permanent total-disability, temporary total-disability, death benefits, and vocational rehabilitation.

Workers' compensation litigation is not like litigation in civil court. On the positive side, proceedings are greatly simplified. Compensation hearings are more informal than court. The rules of evidence don't apply, so lawyers do not play as many procedural tricks as they are wont to do in real courtrooms. There is much more opportunity for a free exchange of ideas between opposing lawyers and the "judge." (There is a wide variation among the states in the title of the hearing officer in workers' compensation proceedings. For convenience I will call this person the "judge.")

There are many features of workers' compensation litigation that are not so positive, particularly in terms of workers' rights. "Judges" in workers' compensation cases are not real judges. Typically they are underpaid and overworked. In some states, like California, they are civil servants, fairly low in the state bureaucratic hierarchy. In other states, like Massachusetts, workers' compensation cases are heard by members of the Industrial Accident Board, made up of political appointees. In most states, the quality of workers' compensation judges in terms of legal education and experience is considerably below that of trial-court judges.

The greatest curse of workers' compensation litigation is the lengthy delay built into the system in most states. The main reason for the delays is there are not enough "judges" and support staff. It costs the state money to provide a litigation system for workers' compensation, and most states would rather spend the money on something else.

In California there are eighty-seven fewer "judges" than are needed, according to the conservative state Department of Finance guidelines. There aren't even enough secretaries and typists. After a California workers' compensation "judge" makes a decision, it may take up to three months before the decision is typed and mailed out. In Indiana the "judges" who decide workers' compensation cases don't even work full time. They are lawyers who maintain their own practices while hearing workers' compensation cases on a part-time basis.

Another reason there is so much delay in this type of litigation is that the very relaxed nature of the procedures compared to real court induces a kind of laissez-faire attitude on the part of the "judges." If an issue cannot be decided because an employer or insurance company has not bothered to produce the appropriate evidence, the "judge" will just schedule another hearing. In Dorothy Hanna's case, the first hearing was requested by her lawyer because Bethlehem Steel had cut off her temporary total-disability benefits. Bethlehem Steel's lawyer got on the phone to the workers' compensation "judge" who had scheduled a hearing and asked him to postpone it for a few months. He complied cheerfully and notified Dorothy's lawyer, Dave Hollenbeck, by mail. Dorothy had to wait five months to get her first day in court.

Delay is usually the friend of the compensation carrier and the enemy of the injured worker. The longer a company can hold on to its money, the higher the margin of profit. For workers, the delay means a permanent loss. The amount of compensation that has to be paid is set in most states at the date of the accident or last exposure to a disease-producing substance. The longer it takes to get the money from the insurance carrier, the less the money is worth. There is no incentive for a carrier to settle a case earlier than is absolutely necessary, and there is often nothing an injured worker can do to speed up the litigation process. Since delays in obtaining a hearing may run from six weeks to a year, and the hearing may prove a waste of time, just another delay until another hearing, the injured worker is strung along from hope to hope and finally to despair.

In some states delay appears to hurt the insurance carrier more than the injured worker. In "agreement" states like Massachusetts and Maine, once a carrier has accepted a claim and started to pay benefits it may not stop them until it has been released from its obligation to pay by the state agency. This requires a hearing. However, the carrier is not obligated to start paying until there has been a hearing. Workers' compensation insurance carriers in these states have a strong incentive to deny claims until they are certain they're going to have to pay.

The workers' compensation system in most states gives the insurance carriers and employers particularly sinister incentive to delay in very serious cases. The worker who has contracted a potentially fatal disease costs the employer or carrier much less dead than alive. Death benefits are so low in most states that it is very tempting for an insurance company to try to delay a case until the worker dies. If John Vandyne, the asbestos worker, lives, his employer, Johns Manville, self-insured, will have to pay a very large medical bill, retroactive temporary-disability, and, probably, permanent total-disability. All this is likely to add up to much more than the relevant maximum death benefit in California, which is $55,000. Furthermore, the death benefit in California, and in most other states, is paid only to "dependents" of the dead worker, not just next of kin. If Johns Manville manages to delay the litigation of Vandyne's case until he dies, it may yet fight a death case against his widow, which could also be delayed several years. Johns Manville could assert that since she was working full time she was not a dependent.

Delays caused by the creaky workers' compensation system destroy workers' lives as surely as low benefits. Unable to plan for the future, in unrealistic expectation of a lump sum to pay off all the accumulated debts and start a new life, the worker waits for the next hearing with mingled hope and dread. While the case takes years to settle by compromise or go to trial, the worker is in a kind of psychological limbo, unsure about the future and unable to go back to work.

Workers forced to litigate a compensation case face more than overworked "judges" and chronic delays. They also face the difficult, sometimes impossible task of finding a lawyer who can adequately represent them. In most states, workers' compensation has evolved into a kind of legal ghetto, with rules and practices all its own, unknown to outsiders. Few lawyers are able to master its complexities without a good deal of practice in the field. Workers' compensation is not taught in many law schools, and where it is taught it is generally an optional night course given by a nonfaculty member and attended by a few of the more zealously labor-minded law students. The best compensation lawyers are those who specialize in that practice. Most of them have loose arrangements with unions that refer to them members who have been hurt on the job.

Nonunion workers often find themselves in the hands of lawyers who know nothing about workers' compensation litigation.

Most states require compensation for a disabled worker to be set at 66⅔% of his or her preinjury wages. This is a remnant of one of the original principles of workers' compensation. It is only the appearance of a connection between compensation and wage loss, however. There's a catch in most states. There is a lid on the total amount that can be paid to an injured worker. Many states prescribe a maximum amount of workers' compensation benefits.

Indiana, for example, has a maximum of $75 per week payable for no more than 500 weeks (not quite ten years). Dorothy Hanna, who was injured in a steel mill in Indiana, had better recover completely in less than ten years. She probably will, but one wonders about other less hardy people who are injured on the job.

Some states make no distinction between permanent partial- and permanent total-disability as far as maximum benefits are concerned. Texas allows a disabled worker to draw workers' compensation benefits for no more than 401 weeks.

States making a distinction between permanent partial- and permanent total-disability often have a maximum amount for partial while continuing total disability for the lifetime of the worker. Thus, Indiana's 500-week maximum is graciously lifted for the totally disabled worker, who is entitled to compensation for an indefinite period. But this liberal provision for total disability disappears when an injured worker tries to grasp it. It's almost impossible to be declared totally disabled in Indiana. No more than ten people achieve this unenviable status each year. "Total disability" in workers' compensation parlance does not mean inability to do your usual job. It means inability to do *any* job, regardless of qualifications. In most states, disability is measured by physical or mental disability. Age, education, and experience are not considered. Thus, in most states a laborer with little formal education and no skills is not considered totally disabled even though he can no longer do any lifting or carrying or any other type of manual work, if he can still do some kind of work sitting down. The fact that no job exists or is likely to exist for this person is irrelevant. He is only "partially" disabled.

What happens to this laborer when his compensation runs out? If he is lucky, he qualifies for and gets disability benefits from the Social Security Administration. It is quite common for those who are deemed partially disabled by workers' compensation to be actually totally disabled. Social Security takes into account the age, education, and experience of the disabled worker and considers whether there is an actual job for him somewhere. Thus, the injured worker who has no chance of getting lifetime workers' compensation benefits, even though actually totally disabled due to an injury or illness at work, may be able to survive with at least a minimum of dignity on Social Security.

Social Security's disability benefits are particularly important to those suffering from an occupational disease. While workers' compensation requires proof that the disability was caused by exposure at work, Social Security requires proof only that the worker is actually unable to work due to an injury or disease, regardless of causation. It's no wonder that of the 581,000 Americans disabled due to occupational disease as of 1980, 44% were supported by Social Security while only 3% were supported by workers' compensation.

John Vandyne is an example of a worker who, disabled on the job by an industrial disease—asbestosis—has not been able to get workers' compensation. He manages on Social Security and his wife's income.

For those workers unable to work due to an on-the-job injury or disease who do not qualify for Social Security, there is welfare in all its depressing manifestations. Aid to Families with Dependent Children (AFDC or just ADC) is available to workers' families on proof of poverty, while county general assistance is sometimes available to workers without minor children.

If an injured worker is on Social Security or welfare, the medical bills may be paid by Medicaid, a federally funded program for medical relief.

It would be a hardhearted citizen indeed who would say that an injured worker, one who has given his health in the production of society's needs, should not be supported by public money in his or her hour of need. The alternative for these workers would be starvation, begging in the streets.

But why should the worker in Texas who has been injured in his or her employer's service be cast out after 401 weeks of workers' compensation? Why should the Indiana employer be responsible for a maimed worker for only 500 weeks? Why should the economic burden of industrial accident and disease fall hardest on injured workers and their families and the taxpaying public?

There are many petty harassments faced by workers who have managed to get workers' compensation benefits. Little tricks of state laws take away covertly what was given overtly. For example, in Indiana, a worker who has been injured on the job has the right to get full medical treatment at the employers' expense. The catch is that the employer has the right to choose the doctor.

This is not an unusual provision. Only recently have many states given the injured worker the right to choose his or her own doctor. California changed its law in 1975 to allow this, and other states are following the trend. But there still are many that vest the prerogative in the employer. The implications are appalling. A doctor who obtains all his referrals from a large factory will not be eager to annoy the source of his business—the management of the factory. The injured worker thus has to go to a doctor who is beholden to his or her employer. Two nasty consequences flow from this. First, the worker receives the type and degree of treatment that the *employer* deems necessary and sufficient. Workers who are "patients" of industrial clinics are often treated like items on the assembly line, items to be patched up in the most cost-efficient way consistent with their relatively low value to the company.

The other consequence of having an employer-designated doctor has to do with workers' compensation. The job-relatedness of an injury or illness and its severity are determined mostly by a worker's doctor. Short of all-out litigation, the injured worker gets workers' compensation based totally on the treating doctor's diagnosis and opinion as to the degree of disability. In a workers' compensation case the injured worker is the plaintiff and the employer is the defendant. If the defendant has the exclusive right to choose the plaintiff's doctor, it seems likely that the doctor will minimize the defendant's (his employer's) liability.

Indiana adds one more little fillip to this favoritism toward the employer. The employer need provide medical treatment only in the county where the accident happened. This provision (which probably violates the constitutional right to travel, but has never been challenged) means that an injured worker must stay under the employer's thumb even if he or she needs to move away. What happens to the worker injured in Hammond who becomes destitute because workers' compensation is so inadequate and has to move in with her parents in, say, Indianapolis? The employer has no obligation under state law to provide medical treatment to that worker.

Another little trick of state workers'-compensation laws in most states is that there is no cost-of-living raise for those on permanent total-disability. Here again injured workers fare worse than people on welfare or Social Security.

Among the plethora of tricks that have grown up in state laws over the years to keep workers from receiving compensation is the requirement found in many state laws that the disabling condition be caused by an "accident." The worker in North Carolina, for example, who picks up a piano and then falls down the stairs with it will be compensated for his injuries, while the machinist whose back finally gives out after ten years of lifting eighty-pound metal tubes is not entitled to compensation. If the injury occurs as a result of the normal duties of the job, it is not an "accident" and therefore not compensable. No medical treatment, nothing. The employee also may not sue the employer for negligence—for example, by recklessly endangering the employee. Workers' compensation is the "exclusive remedy" even if it's not a legal remedy at all.

Recapping the six original goals of workers' compensation, it seems clear that the system has miserably failed the American worker.

A Workers' Compensation Report Card

1. Goal: Internalize the costs of industrial accidents and diseases in the industry that caused them.

   Grade: *F*. The costs of those injuries and diseases that are paid for by workers' compensation are not internalized simply because they are insured and insurance means pooling risks.

Premiums do not reflect actual risk in each occupation and industry as much as they reflect an average risk on a statewide basis. Even including self-insurance, the costs of industrial injury and disease have been largely socialized rather than internalized within industry. In 1974, of the 220,000 Americans who were severely disabled on the job, Social Security supported 47%, welfare 16%, and workers' compensation only 9%.

2. Goal: Compensate workers without regard to fault in order to eliminate the need for litigation.
   Grade: *F*. While the issue of fault in causing an accident or disease has been eliminated by workers' compensation, litigation is rife throughout the system. The issues of job-relatedness and nature and extent of disability have taken the place of fault as reasons for litigation and have proven just as much a source of strife.

3. Goal: Provide medical care for workers' on-the-job injuries.
   Grade: *C*. In cases of traumatic injury, workers' compensation insurers usually provide adequate medical treatment. There are two major drawbacks to this type of medical care, however. In some states, workers do not have the right to choose their own doctors. Secondly, medical care under workers' compensation is available only when an injury or illness has been accepted as work-related. Since work-relatedness is often the main issue in litigation, especially in occupational-disease cases, this means that medical care is not provided until after the case is litigated, which is just the situation that workers' com-

pensation was meant to remedy. In any case, provision of medical treatment is no longer a benefit unique to workers' compensation, as it was in the early part of the century. Now almost every worker is covered by group medical insurance whose benefits are payable without regard to fault or work-relatedness of the condition.

4. Goal: Provide wage-loss replacement for workers disabled on the job.
   Grade: *F*. Few states provide wage-loss replacement for either the partially or the totally disabled. In most states an artificially low dollar amount is set for loss of function of a part of the body and no attempt is made to even approximate wage-loss replacement.

5. Goal: Provide swift and certain compensation.
   Grade: *F*. Workers' compensation has become a painfully time-consuming process for millions of workers disabled on the job. The more serious the injury or illness, the slower and less certain is the compensation. Compensation is not certain in any disputed case. Occupational disease cases in particular are very slow and highly uncertain.

6. Goal: Limit the liability of employers.
   Grade: A. In this one area, workers' compensation has fully lived up to its original goal. Employers are totally immune from negligence lawsuits by their employees.

6

Living the Nightmare

Whether workers' compensation is worth preserving in its present form depends on its ability to adequately help injured workers. Few outside the insurance industry would attempt to argue that it aids victims of occupational disease. As we have seen, no more than 5% of these people receive compensation. But if the system works for victims of industrial *accidents*, there is hope that it can be modified in relatively minor ways to accommodate the occupational-disease problem. If workers' compensation does not function for the vast majority of injured workers, then major reforms are clearly indicated.

The following account of a fictional worker I've called Don Hanks shows how workers' compensation actually functions. His story is a composite of case histories taken from injured workers across the country. I've purposely created a rather mundane character in Don and his wife, Carol, and a routine kind of injury. It might be argued that workers' compensation cannot possibly be designed to compensate something as weird as the male worker who grew breasts because of exposure to the hormone DES. But no one can maintain that it should not adequately and routinely take care of the kind of injury and resulting disability Don Hanks suffered in the following account.

Don lived in a pleasant suburb of a large city. He was a thirty-five-year-old nonunion carpenter making about $9 per hour. He often worked overtime. He managed to provide a decent home for his wife and three small children, although the inflation of recent years had reduced his family's standard of living somewhat. Carol Hanks was thinking of going to work. The children were now all in school and the family probably needed her earning ability more than her presence at home. But not quite yet, she thought.

The day before Thanksgiving in 1978, Don ripped his right hand open while using a power saw at work. He had been operating a saw from which the guard had been taken off. No one at Don's worksite knew who had taken it off. Everyone there used the saw, and work went faster without the guard. Don was driven to the nearest emergency room and was operated on immediately.

Don was in the hospital for about a week. When he went home he had a cast on his right arm. There was a certain amount of pain, but he had some pills if it got really bad or if he couldn't sleep. In any case, he was the kind of macho guy who wouldn't complain about a little pain.

Don called his employer, Reynolds Construction Co., the day he got home from the hospital. He asked Nancy, who was the office manager for this small company, when he would get his workers' compensation check. She was surprised that he hadn't yet received it, and told Don to call the company's workers' compensation insurer. Nancy gave him the phone number and wished him a speedy recovery.

Don immediately called the insurance company. A switchboard operator took his name and phone number and promised that a claims adjuster would call him back. Two days later, he did get a call from the insurance company. A woman named Janet told him that he would receive his first check in the mail a week from the following Friday.

Like most working Americans, the Hanks lived from paycheck to paycheck. Don and Carol were relieved to hear that the check was coming soon and on the same day Don's paycheck would have been due if he had been able to work. Don went down to Reynolds Construction and got the check for the three days he had worked during the previous pay period, before he hurt himself. The company paid him for the whole day of the day he had been injured even though he hurt his hand in the morning.

The check from the insurance company didn't come when it was supposed to. Nor the next day, a Saturday. But it arrived the following Monday. Carol snatched the envelope from the mailbox with a small gasp of relief. Her happiness turned to dismay when she opened the envelope and saw a check for $175. They had been expecting something closer to Don's biweekly paycheck of $600. Don called Janet to ask her what had delayed the rest of his workers' compensation. Janet was friendly, but she told him that the $175 was all he was entitled to. There was a seven-day waiting period, for which Don would get nothing unless he was out of work for at least a month, so he was entitled to only one weeks' payment of temporary total-disability.

Don and Carol sat down together after the children had gone to bed and talked about what to do. Don felt pretty confident that he'd be able to go back to work soon. In the meantime they would just have to try to live on the $175 per week. They'd try to cut their expenses to the bone and not pay some of their bills for a while. They knew, or thought they knew, that when the dust had settled, Don would be entitled to a sum of money for the accident that would enable them to pay the bills that they would now let accumulate.

Still, they would have to borrow money. Their house payments were $300 per month (they were lucky enough to have bought when interest rates on mortgages were below 10%), car payments were $150 per month, and oil to heat the house was now costing the Hanks $300 per month. Just these basic expenses ate up all of Don's approximately $770 per month workers' compensation temporary total-disability payments. There was nothing left over for food, let alone clothes or any other necessities. Luckily, Carol's parents had a little money they could lend the Hanks until everything was straightened out.

Don's main occupation was to get well, and he wished that there were more he could do to actively speed up the process. Every time he went to his doctor, Dr. Wain, he asked when he'd be able to go back to work. Dr. Wain, to whom Don had been referred by Reynolds Construction, would only reply, "Soon. I'll tell you when." After a month the cast came off and was replaced by a bandage.

Don was still unable to do anything with his dominant hand, and it continued to ache quite a bit.

In early February, Dr. Wain told Don that he could go back to work as long as he did no heavy lifting, or pulling with his right arm. Don went back to the offices of Reynolds and showed his former foreman, Bud, the note that Dr. Wain had given him. The note said that Don was released for "restricted duty." Bud told Don that he was very sorry, but there was no such thing as "restricted duty" at Reynolds. Everyone there had to be able to do everything. As soon as Don was able to do his full job, he could come back to work. The general manager affirmed what Bud had told Don and went a little further. He said that he hated to do this but he just couldn't hold the job for Don any longer. He'd have to replace him right away with someone who was uninjured. Of course Don could come back when he got better, if there was a job available then. The general manager wished Don the best of luck.

On his way home Don stopped at his favorite tavern and had a few beers. It turned into a long evening and he got very drunk, which he hadn't done since before the children were born. When he got home he and Carol had the worst argument of their marriage. Don felt like killing somebody, or maybe killing himself. The next morning he told Carol that he'd been fired. She cried at first, then said, "OK. I'm going to get a job. With that and your workers' compensation, we'll get by until you get a job."

On the following Friday, when Carol opened the mail there was a form letter from the insurance company instead of a check. The letter said that since Don was now able to return to work, his workers' compensation eligibility was over and his benefits would cease.

Don was not the type to take all this lying down. First he applied for unemployment compensation. It was less per week than workers' compensation, but it was something. He couldn't believe it when he was curtly told by a man at the state unemployment office that he was not eligible because he had to be able to work immediately without restrictions in order to be eligible. Don protested: "But they told me that I couldn't get workers' compensation because I was ready to go back to work. Now you tell me that I can't get unemployment because I'm not ready to go back to work. What the hell is going on?" The unemployment representative standing behind the counter ignored Don completely and said loudly, "Next." There was a long line of people waiting, and the woman who was immediately behind him rushed up to the counter, practically shouldering him aside.

Don left the unemployment line with a feeling of humiliation that he had never experienced before. He had never received any type of government benefits before; he'd always been able to take care of himself and his responsibilities. No one in his family had ever received any government benefits except Social Security and his dad had been treated at the VA hospital. And now he'd been turned down.

Now Don knew he had to find a lawyer. He wasn't getting his rights, although he didn't know what those rights were. He needed help. But Don didn't know any lawyers and he didn't know how to locate one. Carol remembered that a neighbor had been hit by a car and had sued the driver. She called Beverly, who lived just three doors down the street, and got the name of her lawyer from her. Don called Mr. Johnson, who gave him an appointment.

Don traveled to the city a few days later. Mr. Johnson was a rather pudgy man in a houndstooth suit. Johnson told Don that he would take the case on a contingency basis. The lawyer would receive 20% of what Don eventually got, and if Don got nothing, the lawyer would get nothing. When Don asked Mr. Johnson how he could get his job back, the lawyer said that he couldn't help; that was a matter between Don and Reynolds Construction. Johnson went on to explain that he was a very busy man and that he couldn't talk to Don on the phone or see him again, but that his secretary was trained to handle clients' problems and Don could talk to her if he needed to. The secretary would know when a problem needed to be referred to her boss. Mr. Johnson got up and ushered Don out of the office, smiling affably.

By the time Don arrived home he was caught up in a stewpot of emotions, anger, confusion, and fear. He didn't know where he was at any more than he had before he saw the lawyer. He didn't know what the lawyer was going to do for him or what he could do for him. Don wanted either a job back at Reynolds Construction or his compensation checks. He didn't have either, and he didn't know why he had bothered going to the lawyer.

Life at the Hanks household was becoming decidedly tougher. Carol had found a job as a file clerk, but her take-home pay was only $650 per month. Don had tried to get a job as a carpenter, but his right hand was still weak and no one wanted a carpenter who couldn't pull his own weight. Don didn't know anything else; he had dropped out of high school after the tenth grade. Carol didn't feel that she could ask her parents for any more money, and there was no one else they were close to who could lend them the money they needed. So Don and Carol went to a finance company and took out a loan for $7000. The finance company took a second mortgage on their house.

Don stayed around the house all day while Carol went to work. He was there when the three girls came back from school, and he tried to help around the house. The spring after the accident he planted a garden, which went slowly since he had the use of only one hand, but he found he could do it, and it was very satisfying. Time hung heavily for Don, and he began to feel lonely and sorry for himself. He and Carol argued a lot more than they ever had, and Don had to admit, to himself at least, that he was getting into the habit of picking fights with Carol. She seemed to have lost the sympathy that she had before, right after the accident. She didn't want to hear Don's complaints anymore, and she was often tired and irritable after work. Their sex life went rapidly downhill after Carol started work. It was actually Don who lost interest first. He was always slightly angry, and he was consumed by self-doubt. Carol was becoming resentful and felt a distance opening up between herself and her husband of fourteen years.

One day in late July, Don got a letter in the mail from his lawyer telling him that he should go to a Dr. Hanson for an examination of his hand. Don called the lawyer and was told by the secretary that the examination was for his workers' compensation case and that it was very important for him to not miss the appointment. The doctor spent about five minutes examining his hand. Don was given a grip test. He asked the doctor when his hand would finally get back to normal, but got no response. Dr. Hanson merely shrugged his shoulders and finished looking at Don's right hand. As he examined the hand he dictated a report into a microphone on the wall. Because he was examining and dictating at the same time, the doctor never looked into Don's face, and Don is not sure that he ever saw the doctor's face, except for the instant when the doctor walked into the examining room.

About a month after this exam, Don got a letter from the workers' compensation insurance company telling him that he had to go to a Dr. Randolph for an examination of his hand. Don called his lawyer and was again told by the secretary that it was very important that he go to the exam. The insurance company had the right to have Don's hand examined by the doctor of its choice.

The examination by Dr. Randolph was a repeat of the previous exam except that this doctor was friendlier, and he engaged Don in a little conversation. He asked Don what he did for recreation and was very interested in Don's gardening. Don was a little puzzled when Dr. Randolph asked him if he bowled regularly, since he found it impossible to hold a bowling ball with his right hand.

When the workers' compensation hearing finally took place, it was a big disappointment. Almost a hundred workers and their lawyers were milling around in the dingy halls of a state office building. Waiting with their wives or husbands, or alone, the workers looked worried, or angry, the lawyers impatient and harried. Mr. Johnson rushed up to Don and told him to go into Room D, that he would join Don later. Johnson had several other clients that morning. As Don walked down the hall he saw a series of windowless, crowded rooms, workers' compensation hearing rooms, where "judges" sat listening to case after case.

Don sat down in Room D and waited for his lawyer. He was confused by the chaos in the hearing room. Lawyers elbowed each other out of the way to get their cases heard next; people walked in and out of the small room for no apparent reason. Sometimes the noise from the hallway got so loud that witnesses and lawyers couldn't be heard above the din. The "judge" would then ask someone to close the door. But in a few minutes someone else would walk in and leave the door open again.

Mr. Johnson finally came in and pushed his way to the lawyers' table. He said, "Sit over there," to Don, pointing to a chair near him. These were the only words addressed to Don in the hearing room that morning. After he sat down Don felt that he was invisible. No one seemed to notice that he was there. Johnson, the "judge," and another man had a conversation that seemed totally unrelated to Don. Don had no idea who the other man was. It was the lawyer from the workers' compensation insurance company, but Don never found that out. In five minutes the hearing was over. Don followed Johnson into the hallway.

Standing next to a dirty water fountain, Johnson told Don that the two doctors, Dr. Hanson and Dr. Randolph, had a serious disagreement about the extent of Don's disability. Randolph, the insurance company doctor, had written in his report that Don had admitted to having done considerable gardening, which in the doctor's opinion must have been done to some extent with the right hand. This was a good indication that there was minimal disability in that hand, although the grip test showed otherwise. The doctor concluded that Don was trying to fake the grip test, to exaggerate the extent of his disability.

Dr. Hanson had concluded that Don was totally unable to use his right hand for any productive work. Don's lawyer and the insurance company lawyer had decided, under pressure from the workers' compensation "judge," to agree to send Don to a third doctor. This doctor's opinion would be accepted as binding by both sides.

Unfortunately, the agreed-upon doctor was so busy, and so important, that it took Don six months to get an appointment. He would have to wait until the following May. It was now November, almost one year since the accident. When he heard this news, Don wasn't shocked. He was already feeling a strong sense of unreality; he felt that he was living a nightmare.

Don went home from the hearing dejected. He felt very tired. When the girls came home from school they found him in bed staring at the ceiling. The oldest one, Diane, who was eleven, tried to cheer him up. She'd always been his favorite, and she managed to get him out of bed. He even cracked a smile when she told a joke he'd heard twenty times before.

Again Don pounded the pavement looking for a job. He was willing to take almost anything at this point. He got a job pumping gas at the service station where he had worked when he was a teenager. The owner thought it was odd that Don would want this minimum-wage job, but he knew and trusted him and he took Don on. The owner of the station knew about Don's hand, but thought that he could do the job anyway.

Don and Carol were now in serious financial trouble. Added to their $300 per month mortgage payment was the $165 payment on the second mortgage. And then there was the money owed to Carol's parents. And the bills that hadn't been paid in a year. Dr. Wain told Don just before Christmas that his hand was as healed as it would ever be. It became clear that Don would never be a carpenter again. The Hanks' combined income was now less than Don's take-home pay had been before the accident, and they had accumulated many debts while he was out of work.

Don's final workers' compensation hearing finally took place two years and eight months after the accident. By this time Don and Carol knew not to expect anything; they tried to live each day as it came. Three and a half months after the hearing, Don got a notice in the mail that he was entitled to a permanent partial-disability rating of 21% for the loss of his hand. When Don finally got through to his lawyer to find out what this meant, he was told that his 21% was worth $5285. Out of this the lawyer's fee was $785. Don would get $4500 paid out at the rate of $70 per week for the next fifty-seven weeks. If he needed medical care for his hand in the future, however, he could always have the workers' compensation carrier pay for it.

This composite case has been made up to illustrate some important points about workers' compensation. The story of the wasting of an American worker is not just fiction, however. It is a tale repeated daily all over the country with real people, and results often far more serious than those felt by Don and Carol.

I've spared the reader some of the grotesque consequences of an injured worker's encounter with workers' compensation. Don and Carol did not get divorced, although their marriage went through some pretty rocky times. They did not lose their house, although it took years of financial struggle to pay off their bills. But injured workers do divorce because the strain of trying to cope with the problems created by an industrial accident have proven to be more than the marriage can stand. And they do lose the houses they worked so hard to acquire.

Let's take Don's story and examine it in detail. It should serve as a guide to the problems of workers' compensation seen from the injured worker's point of view.

Don's accident was typical. It was caused in large part by a saw that was being operated without a guard. Who was responsible for this hazard? Workers' compensation doesn't care. It's based on no-fault principles. Don doesn't have to prove negligence in order to get benefits.

This seeming generosity is an illusion, however. If a negligence suit were not barred by the "exclusive remedy" of workers' compensation, the employer could have been held liable for not maintaining a safe workplace. While the early-twentieth-century employer might have been able to assert defenses such as "contributory negligence" or "assumption of the risk," the employer in the 1980s is not so protected. Tort law has changed as much as everything else in our society since 1910. If Don had been able to sue under a negligence theory, he could have recovered all his medical expenses, his loss of earnings (both past and future) as a result of the accident, and a sum for pain and suffering.

After Don returned from the hopsital, he called his employer to find out where his compensation checks were. He was told to call the workers' compensation carrier. There was no further need for the company to pay the slightest attention to its former employee.

The company suffered no financial loss of any kind due to Don's accident. Its workers' compensation premiums were based on years of experience with carpenters. Accidents like Don's had already been factored into the rate. Don's problems were now a matter for the insurance company's claims department. The relationship that Don had built up over the years with his fellow workers and with the foremen and managers of his employer were now useless to him. Don had to deal with a faceless voice on the telephone who worked for an insurance company that he had never even heard of before he was hurt. Don was lucky in that his claims representative was a pleasant, probably well-meaning woman. Many claims people, particularly those who have been doing the job for a number of years, have become cynical and hard-bitten. They've heard the same hard-luck story over and over again, and the same desperate pleas for help. The claims people who answer injured workers' phone calls are on the lowest rung of the insurance company's status ladder. They have no power to do anything to help an injured worker, sometimes they can't even give correct information, and they often respond to the emotional pressure by shutting down all natural sympathy.

When Don got his first check he was appalled that it was so little money. Janet, the claims representative, told him that there was a waiting period of one week. Most states have waiting periods, a certain number of days after the accident when workers' compensation is not paid. The purpose of the waiting period is to discourage "malingering." In the view of many of the framers of workers' compensation laws, there are too many good-for-nothing employees who are salivating to take advantage of the fabulous no-fault benefits available under workers' compensation, and many methods are needed to guard against their abuse of the system. The waiting period is one of those "safeguards." If Don is out of work due to his job-related injury for more than one month, he will get his compensation for the first week.

The check for $175 that Don got from the insurance carrier represented one week's payment of temporary total-disability. This amount was not merely a provision of the insurance contract between Reynolds Construction and the insurance company. It is the maximum amount allowed by state law for temporary total-disability in Don's state. Theoretically, temporary total-disability, paid immediately after the waiting period is over, represents a percentage of the workers' preinjury income. In most states it is 66⅔%. But there is usually a maximum. Lower-income workers do get the 66⅔% wage-loss replacement, while higher-income workers such as Don end up with a much smaller percentage. Don's gross income averaged $360 per week. Thus the $175-per-week temporary total-disability payments represented less than a 50% wage-loss replacement.

The inflation of the 1970s and 1980s has led to higher wages, but temporary total-disability (known as TTD) maximums, established by state law, have been slow to inch upward in some states. As of 1981, the California maximum TTD payment was $175 per week. In Texas it was only $119 per week, while in Indiana it was $130 per week. Some states are relatively generous: Illinois requires insurers to pay a maximum of $353 per week. While the theoretical percentage of compensation for lost wages is the same in all these states, 66⅔% of earnings, it is the maximum amount that really counts.

Don and Carol thought they could manage at least temporarily, in the expectation that Don would soon be able to go back to work. The real trouble came when Dr. Wain released Don to go back to "restricted duty." It is common and sensible for an injured worker to go back to work gradually. The wounds must be given time to heal, but there comes a time when a person can work while still not totally recovered. Most people want to return to work because they need the money. Also, a gradual return to work gives the worker a chance to see what he or she can in fact do despite the injury.

Don, like millions of other Americans, was unlucky enough to work for a company which had no sympathy for this point of view. There is no law in Don's state, or indeed in any state, that compels an employer to rehire an injured worker. Some companies fire injured workers or rehire them on a case-by-case basis. It usually depends on the value of the particular worker to the company or the availability of a job he or she can do. Other companies have policies that simply eliminate the possiblity of rehiring injured workers. Every worker must be able to do every job. Reynolds Construction had such a policy, and Don was out of luck.

Don might not have fared better even if he had belonged to a union. Some union contracts make it impossible for a young worker who has been injured on the job to return to restricted work. Inflexible seniority rules ensure that only very senior workers get the easier jobs.

Don and Carol recovered from the blow of Don being fired only to find that Don's workers' compensation check had been cut off because the insurance carrier claimed that Don could now work. Temporary total-disability is paid only while the worker is *temporarily* unable to work. Once a doctor determines that the worker's condition will not change, it is considered a "permanent disability." The determination of whether a person is unable to work is a medical question. Whether the worker has his old job back or has been fired or actually cannot physically do his former work is irrelevant.

In some states, a worker who has been released for only restricted duty but who has been fired by his employer should still be entitled to temporary total-disability. This might have been the case in

Don's state. But Don didn't know that his temporary total-disability checks should have continued. Janet, the claims representative, might have known that the carrier should not have cut Don off, but she had her job to protect and she knew that economy was the watchword in the claims department. Insurance company employees are rarely penalized for paying out too little, but they may be disciplined for paying too much.

That Don didn't know his benefits should continue does not make him unusually ignorant. He didn't even know that what he called workers' compensation, his $175-per-week check, was in fact only one type of workers' compensation benefit, the one called temporary total-disability. But how should he know? No one had ever explained any of it to him. Only the professionals involved in the system know anything about it. No one at Reynolds Construction really understood workers' compensation or its benefits except in the vaguest way. Only the largest employers have someone who understands workers' compensation and knows what the benefits are. Few indeed outside of the insurance company itself know when benefits may be stopped or must be continued.

Workers' compensation is unique in that while it is governed down to the smallest detail by state law, it is administered almost entirely by private industry—the insurance business. State agencies are involved only in setting rates for the premiums and in deciding disputes between injured workers and workers' compensation insurance carriers or self-insured employers. If Don tried to call the state agency to find out whether his compensation benefits should have been stopped, he would have found no one at the other end of the line whose job it was to answer the question. State bureaucracies being what they are, Don would not even have been able to find out that no one there could answer his question, and would probably have always had the gnawing doubt that if only he had tried to find the right person in the state agency, he would have been able to solve his problems.

Like most people who have been injured on the job and have just begun a relationship with a workers' compensation insurance carrier, Don didn't argue; he thought they must have been within their rights to cut off his benefits. He couldn't find a job as a carpenter, so he went to the unemployment office to file for benefits. Unemployment, unlike workers' compensation, is administered by the state. But Don was turned down. Being fit to go to work immediately is one of the requirements of unemployment. Don's work restriction made him unfit to return to work immediately and therefore ineligible for unemployment.

Some workers in Don's position, feeling desperate, lie and write in the application for unemployment that they are fit to go to work immediately without restriction. If their workers' compensation case goes to litigation, the carrier will obtain the unemployment application by subpoena. The insurance carrier's lawyer may then wave a copy of the application and cry indignantly that the worker himself believed that he was not disabled, and therefore he should not be eligible for any workers' compensation benefits from at least the day of the application. Caught between workers' compensation and unemployment, workers like Don have been whipsawed. They get neither.

When Don was turned down by unemployment, he realized that he was fighting for his life and needed help, and he went to a lawyer. If Don had worked at a union job, his local might have sent him to a lawyer who specialized in workers' compensation. But his was a nonunion job, and he found Mr. Johnson. Don didn't know that workers' compensation is a highly specialized area of law, and that lawyers who don't specialize in it probably cannot adequately represent their clients. Mr. Johnson took some cases from time to time, but he didn't really know much about the subject. For example, he didn't know that Don's compensation should not have been cut off.

Even if Don had found his way to a lawyer who knew workers' compensation, he might not have gotten his TTD restored. In most states lawyers who specialize in this area must handle large numbers of cases to earn the kind of money lawyers expect to earn. In California, experienced compensation lawyers handle 350 open cases or more. Some handle as many as 600. The emphasis of these law offices is on volume and completing cases so the fee can be collected. Even if one of these lawyers knew that a client's TTD should have been continued, the lawyer might not spend the time necessary to litigate that issue.

No one suggested to Don Hanks that he might be a candidate for vocational rehabilitation. In a few states, California included, workers' compensation insurance carriers are required to provide vocational rehabilitation to qualified injured workers, and to continue to pay TTD while the program is going on. The goal of vocational rehabilitation is to return the injured worker to his preinjury wage level as much as possible. If Don Hanks had lived in California, he might have been given a chance to transfer the skills he had acquired as a carpenter into some other trade that he could perform with his weakened hand.

Since California does not keep statistics about the effectiveness of the program, it is impossible to judge its value. It certainly gives hope to those workers who qualify for it, and the feeling that there is a chance to regain some of what they've lost. Most states have refused to mandate vocational rehabilitation, and in this new and more conservative decade it seems unlikely that they will.

Don finally took a job that he'd had when he was a teenager. It was crushing for him to admit to himself that he had to go back to where he started from, that all his hard-won acquisition of carpentry skills was worthless. But it was worse to be idle. Don was fortunate enough to be able to work. Welfare was still unthinkable for the Hanks, but many workers, injured worse than Don, have had to accept it and its tacit admission of total failure. The real failure, of course, is that of the workers' compensation system to help salvage injured workers' lives.

Don Hanks had some unpleasant and typical experiences with doctors after his on-the-job injury. He was referred to Dr. Wain by his employer. This is generally the way most workers find their doctors. In some states they do not have the right to choose their own treating doctor for an on-the-job accident. The employer has that right. But even in states where there is free choice of doctors, workers often go to ones recommended by their employers.

Don went to two other doctors, Dr. Hanson and Dr. Randolph. They merely examined him for the purposes of litigation. They were not interested in treating him; they did not treat many people. They both made a handsome living by examining workers who had been injured on the job and writing reports that could be used to evaluate disability. There are many doctors who do this kind of work in major metropolitan areas, usually specialists of one type or another. Their reports are used to determine the most hotly litigated issue in workers' compensation: the degree of permanent partial-disability suffered by a worker as a result of an industrial accident. In the case of an industrial disease, the doctor's task is usually to give an opinion on the question of whether the condition is work-related or not as well as the degree of disability.

Unfortunately for Don, the fictional Dr. Hanson and Dr. Randolph were both known as "whores" for their respective sides. Their reports on the condition of Don's hand were diametrically opposed.

Whores' opinions are known to the initiated to be always biased, either in favor of the defense, the insurance carrier, or the injured worker. The opinions of these doctors are like the asking price of goods in the oriental bazaar. You have to discount by half and then start bargaining down.

Dr. Randolph, who was a "defense whore" (on the insurance company's side), was typical of many of his kind. He found out that Don had done some gardening and he used that in his report to discredit him. Randolph asked the question about bowling after he had apparently gained Don's confidence, in order to get more evidence to discredit him. He probably didn't actually think that Don could bowl with his injured hand, but it was worth a try. If the gardening had not turned up, Randolph could have written in his report, "Patient denies using his right hand to bowl."

A doctor's value to a workers' compensation system as it presently exists depends on two factors. The first is his or her understanding of workers' compensation law and procedures. The medical opinion must be couched in terms that are recognized and understood by workers' compensation lawyers and "judges." There are key words in many states and they must be used with precision. In California there is a vast difference between a condition that is diagnosed as "minimal" and one that is "slight." "Minimal" means no compensation. The second factor that must be considered in assessing the usefulness of a doctor to the system of workers' compensation litigation is the credibility of the doctor.

The main function of the "judge" in a case like Don's is to decide the degree of permanent partial-disability. To do this the "judge" needs medical evidence. Doctors cannot decide the case, since "disability" in workers' compensation is a legal concept, but "judges" cannot make a decision without medical evidence to guide them. The "judge" in Don's case knew by reputation and his own experience from previous cases that both doctors who had examined Don were "whores." Nevertheless, the "judge" could have made a decision based on the reports of the two lawyers if he had been so inclined. Many "judges" would simply have split the difference and given Don an award for permanent partial-disability that was halfway between the opinions of the two opposing doctors. This would have had the virtue of speed at least. But the "judge" who heard Don's case was punctilious. He insisted on delaying the case to get better evidence. And he prevailed on the lawyers to agree on another doctor for an evaluation. The doctor they agreed on was one of the small number in the area whose opinions were respected by both insurance lawyers and workers' lawyers. Typically, there are so few of these doctors that it takes many months to get an appointment with one.

By the time Don's case was settled, it was obvious that workers' compensation was irrelevant; it could not help him or his family. When he needed money the most, when he was totally unable to work, he didn't get enough to survive. When he was able to go back to work on a restricted basis, workers' compensation didn't prevent his employer from refusing to take him back. And when he lost his job, workers' compensation cut off his TTD. The carrier did provide medical treatment, but Don could have received that from his group health coverage and later from Carol's group health plan. The workers' compensation litigation seemed to drag on forever—almost three years, as it turned out. And it ended with a whimper. The $4500 Don received as permanent partial-disability made no sense at all. It was grossly disproportionate to the loss of the use of his right hand and his fall from the ranks of the skilled. He had lost in wages much more than that each year since the accident. But the $4500 was to be his only compensation, other than the TTD he had already received. Don could count himself lucky that he didn't live in Texas or some other states where the TTD he had been paid would have been deducted from the permanent partial-disability (PPD), leaving him with nothing.

7

The Insurance Cartel

While the workers' compensation system cannot better the lives of most workers hurt on the job, it is very effective in making rich insurance companies richer. Immune from antitrust laws, the insurance industry has organized itself into a cartel that virtually dictates to employers what they shall pay in workers' compensation insurance and to workers what they shall receive in benefits. From the cradle of rates to the grave of benefits, all is quietly managed by a small group of insurance executives. Operating in welcome obscurity, they accumulate billions of dollars each year.

Workers' compensation is an enormously expensive way to take care of injured workers. Out of every dollar paid in premiums, no more than sixty-five cents is paid out in benefits. In some states almost fifty cents of each premium dollar is retained by the insurers. By contrast, the Social Security Administration uses only one and a half cents for administrative expenses for every dollar received in Social Security taxes.

Table 3 shows just how expensive the workers' compensation system is in California, the largest insurance market in the country. It shows the total amount of premiums paid to insurers by employers in a five-year period and the total amount of money paid out by insurers in benefits to workers for the same period. The amount shown as kept by insurers is for all purposes, including payment of future benefits for claims that arose between 1976 and 1980. The exact amount of this "loss reserve" is determined by the insurance companies themselves. No more than about $2 billion of the reserve will eventually have to be paid out. In the meantime, the money is invested.

Table 3
All California Workers' Compensation Insurers
1976–1980

Total Amount Taken In*	$10,222,583,000
Total Amount Paid Out for Losses	4,785,158,000
Total Amount Kept by Insurers	5,437,425,000

*Earned premiums less dividends paid to policyholders.
Source: "Underwriters Report," Statistical Review Number, May 1981.

Workers' compensation is by far the largest line of commercial insurance, dwarfing all others in premium dollars collected and earned. It is the second-largest line of insurance overall, surpassed only by automobile liability insurance. More than 90% of America's workers are covered by workers' compensation. In 1980 the 600-odd companies that offer workers' compensation earned almost $14 billion in premiums. In the same year it cost over $4½ billion just to pay the overhead for these companies.

Workers' compensation is a combination of private enterprise and strict government control. State governments determine benefits and resolve disputes in state agencies. The rates upon which premiums are based are ostensibly set by state insurance commissioners, and deviations are usually not allowed. Private insurers operate the system, however. They sell policies to employers and collect premiums from them. Private insurers disburse benefits to injured workers with no supervision from the state.

This odd combination of private sector and public is a historical curiosity, a one-of-a-kind benefit program whose form was never copied. When it was instituted, the very idea that government could or should legislate reform was novel and, to some, revolutionary. Leaving the administration of the program to private insurers seemed natural. Workers' compensation was run by private insurers in Great Britain, whose laws provided the model for most states. Except for federal employees, all the workers' compensation action was at the state level. Until the New Deal of the 1930s it was thought that the federal government could not legislate such programs.

Social-benefit programs that originated after workers' compensation were designed to be government-run. Other than workers' compensation, the only major social-benefit program that is run by private industry is Medicare. This program, which is itself in trouble, relies much less on private enterprise than does workers' compensation. Benefits are paid by "fiscal intermediaries" that are conduits for federal money. Workers' compensation, alone among major social-benefit programs in the United States, allows private industry to control both the system of collection of money and distribution of benefits.

Eighteen states, moved by populism or desperation, created their own insurance companies. Six of these closed their borders to private workers'-compensation insurers. Despite many efforts to overturn these state-run systems, Ohio, Washington, West Virginia, and a few other smaller states have kept private insurers out. (More on this later.)

From its inception, workers' compensation was recognized as more than just another form of insurance. It was introduced as a progressive social-benefit program and as such had certain features not found in other types of insurance. It was to be paid for not by the intended beneficiaries, workers, but by their employers. Also, workers' compensation was to be strongly favored. Employers were to be encouraged to insure their workers. These policy considerations resulted in state laws that provide serious penalties for employers who don't insure. They range from prison terms of up to one year, fines of up to $5000, the closing of noncovered businesses, and that most dreaded of penalties, the abrogation of the "exclusive-remedy" provision of the workers' compensation law.

Almost all states allow employers who can qualify to "self-insure"; to pay their own workers' compensation claims without resorting to an insurance company. Only the largest and most financially solvent of businesses qualify for this exemption. Just 1% of American businesses are self-insured for workers' compensation. These are the really large corporations, however, for 15% of American workers are employed by this small group of employers. All other employers must buy workers' compensation insurance.

Even the self-insured are not totally divorced from the insurance industry, however. Most of them hire insurance brokers to adjust claims and litigate all disputes. The worker who is injured while employed by one of these self-insureds will have his or her claim handled completely by the broker and will usually not even know that the employer is really paying for the claim. His employer is insulated from the injured workers' problems almost as much as employers who have bought workers' compensation insurance.

The clumsy blend of private enterprise and public regulation known as workers' compensation is in deep trouble in many states. Premiums are too high for employers and benefits are too low for injured workers. Neither business nor organized labor is happy with the system. Many state legislators are beginning to realize that if something is not done soon to make the system work better, drastic changes will have to be made later.

Workers' compensation has rarely been the subject of controversy before, for a number of reasons. For many years premiums were relatively low, a negligible business expense. Benefits were static, but so were wages and prices. Not many cases were litigated, and workers tended to take what they were offered. "Judges" didn't look kindly at new theories of compensability. These were the halcyon times for workers' compensation insurers. Clayton Jackson, chief lobbyist for California's insurers, told me that this line of insurance was traditionally so certain that insurance people didn't have to think. The system ran itself, and nothing was more certain than the profits from workers' compensation insurance.

The predictable universe of workers' compensation, so entrenched and stable since its beginning, began to rumble with the noises of change in the 1970s. Inflation devalued the old benefit levels that had been written into state law. The formula for compensation of temporary total-disability and in some states for permanent total had been originally established as 66⅔% of the lost wage of the injured worker. Most states had a maximum limit that was intended to prevent the highest-paid workers from getting more than was actually necessary to continue a reasonable standard of living. Inflation drove wages so high that the maximum limits on compensation put a cap on benefits for great numbers of workers. The effect of this was to drive down the percentage of lost wages given in compensation to injured workers. Most workers injured in 1940 got 66⅔% of their lost wages in workers' compensation while they were disabled, but by 1970 most received less than 50%. In California in 1970 the maximum weekly payment for temporary- or permanent total-disability was $52.

In the early 1970s benefits rose considerably, but not in all categories of workers' compensation. Temporary total-disability has shown the greatest increase: in California the maximum in 1981 was $175. This is still inadequate, and it is still not 66⅔% of lost wages for many California workers, but there has been some movement. Permanent partial-disability, which represents 60% of all cash benefits paid as workers' compensation, is another story. In California as in many other states, there have been no changes in the schedules that dictate permanent partial-disability awards. As of 1981, the loss of a leg at the hip in California is worth only $32,000. In Indiana the insurer has to pay only $16,000 for it. The cheapest of all is Massachusetts, where a worker who loses a leg at the hip can expect a maximum of $6000 for his or her loss.

While benefits stagnate, rates, the basis of premiums, have been tacitly indexed to inflation. In state after state the cartel of insurers comes back to the insurance commissioner each year with a new request for rate increases—and the requests are routinely approved. But the insurers may be overreaching themselves. Businesses that must pay the premiums are becoming alarmed at the cost of workers' compensation insurance. The public is starting to wonder what the insurers do with the vast amounts of money they collect. Questions are being asked about the incredible profitability of workers' compensation insurance resulting from high returns on investments.

The old comfortable way of doing workers' compensation insurance business is being challenged by new realities. Insurers are facing problems of unprecedented complexity. These problems threaten to overturn the whole system if not solved adequately. Inflation, and especially occupational disease, threaten to shake workers' compensation to its foundation, and perhaps to topple it altogether.

Occupational Disease

Occupational disease is the insurance industry's bugaboo. The dark revelation that more than 100,000 Americans are dying each year from this cause has frightened insurance executives. They are apprehensive about the increasing awareness of the contribution of the work environment to the development of many diseases. Substances that were regarded as safe are now known to cause cancer. Heart disease, once thought to be strictly non-work-related, is now being linked to stresses at work. If workers' compensation insurers had to pay for any large percentage of the appalling toll of occupational disease, their margins of profit would be seriously affected.

Since no more than 5% of those suffering from occupational disease now receive workers' compensation for their condition, the burden of caring for these disabled men and women has fallen on the public. Social Security, welfare, food stamps, Medicare, and Medicaid take care of the people who should be the workers' compensation insurance industry's charge. In 1980 a total of $1.9 billion, or 13% of the Social Security Disability Insurance (SSDI) Trust Fund of $15 billion, was paid to victims of occupational disease. For the same year, SSDI paid out $500 million for medical care for job-related diseases. Thus, the Social Security system put out $2.4 billion for occupational disease in 1980. This represents money that the workers' compensation insurance industry did not have to pay, although it collects premiums and receives the protection of law under the theory that it and it alone is the appropriate mechanism for compensating workers who have been injured or made sick on the job.

As yet, few occupational-disease claims are filed; fewer still result in compensation. This is OK with insurance industry leaders. Their response to the problem of occupational disease is to deny its existence.

The insurance industry point of view on occupational disease was summed up by John A. Antonakes, vice-president of Liberty Mutual, the largest workers' compensation insurer in the United States. At a seminar on occupational disease sponsored by the National Council on Compensation Insurance, Antonakes said that there were three categories of occupational disease.

1. The commonly accepted occupational diseases such as metal poisoning, silicosis, asbestosis, byssinosis, etc.
2. Ordinary diseases of life such as asthma, emphysema, bronchitis, hearing loss, heart disease, emotional illness.
3. Cumulative injury.

He says that claims for the first category amount to no more than 3% of all workers' compensation claims and are not expected to increase. As for the second and third categories, if all the potential claims in those groups were compensated, "the costs could be staggering." But in his view, these diseases and resulting disability should not be taken care of by workers' compensation.

The Liberty Mutual executive illustrated his points with the example of byssinosis due to cotton-dust exposure. The problem of this occupational disease, like others, is vastly overstated. He says that the total worker population exposed to cotton dust is 35,000. Of these, about 3500 are at risk of suffering byssinosis to the extent that their working capacity will be affected. Antonakes presented these figures to an audience that wanted to believe him, fellow insurance executives mainly. His numbers are in glaring contrast to those maintained by the federal government's National Institute of Occupational Safety and Health. NIOSH estimates that 300,000 workers are significantly exposed to cotton dust and many more are potentially at risk. At least 35,000 workers are permanently and totally disabled due to byssinosis, and 800 more become totally disabled each year.

As for the "ordinary diseases of life" and cumulative injury, Antonakes implies that these should not be compensable. He warned that "If public policy requires that compensation be provided to diseased workers whose employment had only a peripheral influence over the disease itself, then we would have serious doubts as to the capacity of the workers' compensation system as presently structured to handle the potential flood of claims."

The key word in this statement is "peripheral." Antonakes is raising the issue of "arising out of and in the course of employment," the "AOE/COE" test that is the main issue in litigation of occupational-disease cases. The public-policy issue is whether workers' compensation should cover diseases that are partly caused by work—should such diseases be deemed AOE/COE? Antonakes believes, like almost all insurance people, that they should not be compensable. The worker who smokes whose probability of developing lung cancer increased fourfold by working around asbestos would not agree.

Antonakes expresses the point of view of most workers' compensation insurance executives, content with the old way of doing things and not wanting to change to keep pace with a changing world. The absurdly low figures of people at risk of byssinosis, the firm refusal to consider modifying the compensation system to take care of "ordinary diseases of life" that are actually caused by work, these are signs of a kind of denial that strangely parallels the denial of workers who have been exposed to occupational-health hazards. Hoping for the best, compensation insurers take comfort from the fact that past occupational-disease claims have been few and the present rules of compensation litigation make it very difficult for diseased workers to win a claim.

Antonakes' comfortable and self-serving view of the occupational-disease question is implicitly refuted by the leading expert on workers' compensation as it pertains to occupational disease. Peter Barth, in his *Workers' Compensation and Work-Related Illnesses and Disease*, says that the estimates of 100,000 occupational-disease-caused deaths per year may be too high or too low. * No one knows what a really accurate number would be, because there is no agreement about the definition of occupational disease. But, he says, "in absolute terms the occupational-disease problem is a large one," and "the evidence suggests that the workers' compensation system in the United States is dealing with very few cases of occupational disease."

Another of the small group of experts in workers' compensation believes that there already is a crisis in the workers' compensation

* Peter S. Barth with H. Allen Hunt. *Workers' Compensation and Work-Related Illnesses and Diseases*. First edition. Cambridge, MA: MIT Press, 1980, p. 255.

system as a result of occupational disease. John F. Burton, Jr., former chairman of the National Commission on State Workmen's Compensation Laws, declared at the same National Council on Compensation Insurance seminar where Antonakes spoke that occupational-disease cases are already inundating workers' compensation. Back strains and sprains, which accounted for 20% of workers' compensation claims in 1980, have traditionally been classified as traumatic injuries rather than occupational disease. But, Burton says, most bad backs are actually caused by disease, rather than by traumatic injury. When workers' compensation case-law developed, back injuries were mainly the result of a trauma and resulted in a herniated disc. Now there are at least three other types of back problem that result in disability and are the subject of workers' compensation claims. These other conditions are found in the general population and are virtually impossible to link to a specific event, such as an accident at work. Burton says, ". . . while it is apparent that a set of legal rules can be superimposed on back cases to distinguish those that are work-related from those that are not, we ought to recognize that from a medical or scientific standpoint those rules are pure hokum."

Burton's conclusion is that in order for workers' compensation to properly compensate occupational disease, it must do away with the work-relatedness test. The AOE/COE issue should simply be eliminated from consideration in occupational-disease cases. In a sense, Burton confirms Antonakes when the latter says that workers' compensation "as it is presently structured" cannot handle the "potential flood of claims" due to occupational disease. To remove the work-relatedness test is to radically change workers' compensation, to restructure it in a basic way. Perhaps that's just what is needed.

Litigation

Most occupational-disease cases are litigated. These claims are usually begun by a worker's lawyer filing an application with the state agency, asking it to adjudicate the claim. Insurance executives point to this circumstance to show that occupational-disease claims are an unwarranted burden on the system. In fact, the reason for the great amount of litigation is that insurance carriers will not pay reasonable compensation without being ordered to by the state agency.

An important reason for carriers' intransigence is that in any particular claim there are usually two or more insurers involved. Workers' compensation coverage, like other insurance coverage, is for limited periods. An employer may have more than one compensation insurer over a period of years, or an employee may work for a number of employers, each having a different insurance carrier. A carrier is responsible to pay workers' compensation only for injuries occurring during the period it covered a particular company and its employees. But occupational disease often takes years to develop. Thus, many insurers may potentially be responsible for a single claim. Insurance carriers make it a point of honor never to pay any part of a claim that may be the responsibility of another carrier. Injured workers come and go but competitors are always there. Reputations of claims managers and insurance company lawyers are made or lost depending on cases that involve other carriers. A good reputation, given the values of the industry, means you don't pay out a dime more than you absolutely have to. It is common for litigation to go on for years simply over the issue of which carrier is to pay the claim. All parties may agree that the worker should be compensated, but he or she just has to wait until this issue is finally settled.

The painfully slow nature of workers' compensation litigation works to the benefit of the insurance carrier. Delay is the friend of the carrier. There are several reasons for this.

Insurance companies earn a good part of their income from investing their reserves. The longer they hold on to their money, the richer they get. If a worker is disabled in 1979, but does not get compensation until 1981, not an unusual circumstance, the insurance carrier has had the use of two years' worth of his or her compensation in a high-interest investment market. Also, in most states benefits are determined as of the day of the accident, regardless of when payment is actually made. The carrier that delays payment through litigation is rewarded by being allowed to pay off in dollars that are worth less than they were on the date they should have been paid. Penalties for late payment are usually minuscule and hard for the worker to collect. In California, the penalty for late payment is a 10% augmentation of the particular type of benefit unreasonably delayed. The worker has the burden of proving that the delay was unreasonable. Most workers' compensation lawyers in that state don't even bother to try for this penalty. "Judges" hate to award it—it makes the insurance companies angry.

A more sinister advantage for delay involves death benefits. The worker who has contracted asbestosis or cancer as a result of exposure at work costs less dead than alive. In most states, if a worker can prove his or her condition compensable, the insurance carrier has to pay much more in medical bills and total disability payments than in the death benefits. It is cheaper for a carrier to litigate a case for years, if need be, until the worker dies than to pay while he or she is still alive. Death benefits are payable only to a dependent spouse and children; other next of kin are not entitled to them. If a single man or woman, or a widow or widower or a divorced person with grown children, is killed on the job, the carrier is responsible only for a small burial allowance. These workers are bargain-basement specials for insurers; all they have to pay is a burial allowance. In 1981 in New York this was $1250; in California $1500.

Workers' compensation cases, unlike ordinary negligence cases, do not "survive" the death of the worker in most states. If a person is killed off the job through someone else's negligence, his heirs, whether dependents or not, could collect most of the damages the victim could have collected had he survived. In addition, many states allow suits to be brought by next of kin for "wrongful death." The disparity between workers' compensation, with its paltry death benefits, and the law of negligence as it has developed in this century is a commentary on the legal ghettoization of the American worker in the workers' compensation system.

Workers' compensation was designed as a no-fault system in order to eliminate litigation and to ensure prompt payment of benefits. It was made the "exclusive remedy" on the assumption that it would compensate all work-related injuries and diseases, if not all at the beginning, then eventually. But, as we've just seen, the system is rife with litigation. Injuries and diseases that are caused by work are not being compensated. Delays are robbing injured workers of the value of their benefits as well as hope for a new beginning.

Meanwhile, rates are going up and employers are becoming increasingly restive. Every increase in benefits brings with it an increase in rates, in a seemingly classic example of the inflationary spiral. Yet benefits are still too low. And too few work-related disabilities are being compensated, especially occupational-disease cases. There just doesn't seem to be enough money in the workers' compensation system to take care of all the people who should be taken care of.

When there isn't enough money to run a social-benefit program the way it has been run in the past, there are two basic avenues of change. They are not mutually exclusive, but generally those who are interested in one approach are not interested in the other. One approach is to focus on benefits and attempt to restrict them in some way. The other approach is to change the method of financing to provide more money for current benefits and for the possibility of expansion of benefits in the future if necessary.

The insurance industry has attempted to effect "reform" by restricting benefits. Insurance spokesmen seem to have convinced themselves, and are trying to convince others, that all the problems of workers' compensation are the result of wildly liberal "judges" giving away huge sums of money to undeserving cheats at the insistence of unscrupulous plaintiffs' lawyers. It is this nefarious group of troublemakers, insurers allege, that is plundering the system.

These arguments obscure the underlying reality of workers' compensation. It is a system controlled in most details by the insurers themselves. Workers have only the right to contest a denial of benefits, but insurers hold the money and the power. Employers, particularly small ones, have no rights at all. They must buy compensation insurance at rates prescribed by state law.

Some of the abuses that insurance people talk about do happen. Occasionally someone fakes an injury or a compensation "judge" makes an award of unwarranted liberality. But it is not possible to believe that great numbers of American workers are attempting to freeload off an overgenerous workers' compensation system, just as it is incredible that there are great numbers of too-liberal "judges."

The financial problems of workers' compensation don't really have to do with benefits at all. It's the other end of the system that has to be examined to explain why premiums are so high and benefits so low. It's not the payout to the workers that is causing the crisis, it is the sheer inefficiency of a system that allows insurers to take billions from employers and give relatively little in return. For the truth is not that injured workers are getting too much but that they are getting much too little.

The Cartel

The insurance industry is a cartel in the classic sense. Immune by a special federal law from antitrust laws and federal regulation, it has grown into one of the most sophisticated and powerful economic and political forces in our society. Organized and cooperative, American workers' compensation insurers have come to dominate the governmental forces meant to regulate them.

On the fifty-first floor of a skyscraper overlooking midtown Manhattan are the offices of the cartel's nerve center. The National Council on Compensation Insurance decides what workers' compensation insurance will cost in most of the United States. It has a low profile, and no doubt wishes to keep it that way, for if its real power were known and appreciated, it would no doubt become the subject of public scrutiny, and embarrassing questions might be asked. Governed by committees of representatives of the insurance companies that support it, the National Council makes rates for thirty-two states and provides critical technical assistance to local rate-making organizations in twelve more jurisdictions. More than 1100 people work for the National Council, which provides services to more than 600 member or subscriber insurance companies.

While all government regulation of insurance is done at the state level, and the federal government is forbidden by law to regulate it, the insurance industry is tightly organized on a national basis. Few state governments are a match for the technical expertise of the National Council or the political muscle of the insurance-industry trade groups. When the National Council proposes rates based on national profit and loss data, most state insurance commissioners have little choice but to accept its conclusions, since their resources are pitiful in comparison with those of the National Council.

California and a few other states have different organizations that serve the same function as the National Council and are also creatures of the insurance industry. The California Insurance Rating

Bureau, a semigovernmental body run by private insurers, has a monopoly on recommendation of rates to the state insurance commissioner.

Workers' compensation insurance is compulsory and its rates are noncompetitive, particularly to small business that does not benefit from experience rating or retrospective-rating plans. The rates are set by the industry itself and are rarely challenged by relatively weak state insurance commissioners. And the rates produce billions more each year than is needed to pay claims and operate the system.

There are three types of workers' compensation carriers, two privately owned and one state owned. There are stock companies, mutuals, and state funds. Stock companies and mutuals write other lines of insurance as well as compensation. State funds are unique to workers' compensation.

When workers' compensation was introduced, some states found that private insurers were so rapacious that many employers would be ruined if they had to pay the premiums that were demanded. Eighteen states organized their own insurance companies, called "state funds," to provide insurance at reasonable rates. Of these, six are "exclusive state funds"; in six states only the state fund may sell workers' compensation insurance. The other twelve state funds compete with private insurers and operate for the most part like them. The largest exclusive state funds are in Ohio, Washington, and West Virginia. Of the competitive state funds, California's is by far the largest. State funds are nonprofit, returning their excess to their policyholders. Their degree of efficiency and share of the market vary greatly from state to state, depending essentially on the role determined for them by the state legislature.

Private insurers of all types find it appalling that there are states with exclusive state funds. Every five or ten years they try to bust the monopoly in the more important states. In 1980 they tried to open up Washington and Ohio to private insurance, and they failed in both states. The Washington legislature was hit by a concerted lobbying effort that nearly succeeded in introducing private insurance. The insurance industry bill lost by only one vote.

Ohio is a constant challenge to some insurers. Situated in the heartland of the industrial part of the nation, large and relatively prosperous, Ohio sets an example for the rest of the country, a bad example by insurance industry lights. The Ohio State Fund kept premiums for employers low and benefits for workers high. Both workers and employers were happy with the fund's performance.

There was no chance that the Ohio legislature could be convinced that private workers'-compensation insurers should be allowed to operate in the state. That had been tried no fewer than twelve times before. As recently as 1978 a bill had been introduced but had so little support that it failed to get out of committee. So a group of workers' compensation insurers banded together and tried to buy their way in. The strategy they chose was to initiate a referendum for a constitutional amendment that would have forced the legislature to allow private insurers to compete with the Ohio State Fund.

Opposition to the insurance industry's amendment was almost universal in Ohio. The manufacturers' association, the Chamber of Commerce, and the Ohio AFL-CIO banded together in an unprecedented alliance against the out-of-state insurers. The war chests were unequal. Insurance interests poured over $6 million into a slick direct mail and media campaign run by a public relations firm. Opponents of the amendment spent only $890,000. In an election that showed that money alone can't always buy the voters, the insurance amendment was turned down by a staggering 78%.

In the forty-four states that do not have exclusive state funds, stock companies and mutuals compete for the privilege of insuring employers. Stock companies, like other conventional corporations, are owned by stockholders who are paid dividends, according to the number of shares they own, out of the company's profits. Mutual companies are nominally owned by their policyholders. If there is any profit left over after expenses are paid, it is distributed to the policyholders according to the size of the premiums they paid. There are many more stock companies writing more premium dollars than mutuals, but mutuals make a much higher underwriting profit. Stock companies sell their insurance to employers through brokers, who earn high commissions, while mutuals usually maintain their own sales staff.

There were pitched political battles in the early years of workers' compensation between stock companies and mutuals. The stock companies, which represented the old way of doing things, accused the mutuals of being socialism in disguise. Mutual insurance companies are now very much a part of the establishment; actually the leaders of the workers' compensation insurance industry. There are still some issues in dispute between mutuals and stock companies, however. Minimum rates are now set by insurance commissioners in most states. Mutuals would like to keep this system, while stock companies are pushing open competition, deregulation of rates.

Under the regulated-rate system, large and small businesses are on a roughly equal footing, since they pay the same rate for the same job classification. It is more profitable for insurers to cover large companies, however. Mutuals can attract the business of the large employers because although they must charge the same rates as stock companies, they offer larger rebates in the form of dividends than stock companies are able to offer. If rates are deregulated, the larger stock companies will be able to reduce their rates to compete with the mutuals for the larger employers. Spokesmen for the mutuals argue that the result of deregulation will be higher rates for small employers.

Table 4 shows the 1980 underwriting profit of stock companies compared to mutuals. Mutuals have a much higher profit ratio than stocks, indicating that it is much more profitable to insure large risks than small ones. Deregulation of rates therefore might set off a scramble for the larger employers' insurance business, and result in lower rates for them. It is unclear what the effect of deregulation might be on small employers.

Table 4
1980 Underwriting Experience
Stocks and Mutuals Compared

	Stock Companies	*Mutuals*
Premiums Earned	$10,422,354,000	$3,278,184,000
Underwriting Profit	357,947,000	471,729,000
Ratio of Underwriting Profit to Premiums Earned	3.4	14.4

Source: A. M. Best, *Aggregates and Averages* (insurance periodical), 1981.

Making Rates

The schedule of benefits paid to injured workers in a particular state is only one factor among many that determine what rates shall be. Many other factors come into play, including such political ones as the strength of the insurance lobby in the state. The ratemaking process is the key mystery of workers' compensation. While controversy swirls around the issue of benefits, the insurance industry quietly goes about dictating how much employers will have to pay for insurance and where that money will go. The story of how rates are established sheds a great deal of light on the infrastructure of workers' compensation. If workers' compensation can be called a system, it is because of the uniformity of ratemaking, a uniformity unmatched by any other aspect of the varied pattern of state compensation programs.

Rates are determined by estimates of the amount of money an insurance carrier will have to pay out to cover losses. These estimates are called "pure premiums." Added to them is an amount to cover all the insurers' expenses and allow for a profit. Two and a half percent of the rate is allowed for profit. A separate rate is set for each of over 600 different occupations.

Rates are developed using a combination of two sources of data. The most important source is a combination of the history of figures of past premiums written, premiums earned, losses paid, and losses incurred on a statewide basis without regard for the variations of individual industries. The other is the record of payroll losses incurred for individual industries, which is factored into the rates for the particular industry. The two sources are combined so that the average premium dollar will go 65% for losses and 35% for expenses and profits. Carriers are not overly concerned if the expense loading on any particular rate is too low or too high as long as the overall state rate allows the 65%/35% balance.

Some of the premium money received by carriers from employers is set aside to cover claims not yet paid and possible cancellations of policies. State laws require carriers to establish "unearned premium" reserves and "loss reserves." When a workplace accident happens, a carrier must estimate the amount of money that will eventually have to be paid in benefits and establish a reserve to pay the potential loss. The money in the unearned premium fund and the loss reserve is invested by the carrier.

Rates are published each year in a manual, but not every employer pays the manual rate. Larger employers pay rates that vary on their own loss experience. "Loss experience" means simply the number of work-related accidents or diseases suffered by employees of an industry. Called "experience rating," this variable factor in ratemaking is supposed to provide an economic incentive for employers to clean up their workplaces. "Retrospective rating" plans also provide a reduction or increase in premiums charged depending on the safety record of an employer.

Regardless of safety experience, many large employers pay rates below the manual rate, receiving rebates or dividends on their premiums. Insurers comply with the minimum-rate laws by charging the full premium and then returning a part of the premium. In states where competition is keen, like California, some insurers collect a discounted premium.

Let's look at what employers have to pay and its relationship to benefits actually paid out to injured workers.

A premium is the money an employer has to pay for workers' compensation insurance. The premium is determined by multiplying the rate times the dollar amount of the payroll. For example, if the rate is $2.50 (rates are always expressed in terms of dollars per hundred) and the payroll is $10,000, the workers' compensation premium will be $250. There is a different rate for over 600 occupations. Each state has a different rate for each occupation, and some state rates are much higher than others. Common sense and the insurance carriers tell us that rates will be high when benefits are high and low when benefits are low. Nothing is further from the truth. Statistics show that there is no significant relationship between rates and benefits. Table 5 illustrates this surprising fact.

The table shows four states with relatively high rates and low benefits and four states with low rates and high benefits. In each group of four selected states there are two relatively small states and two large states. Three of the high-rate states are especially stingy with permanent partial-disability benefits, since they deduct amounts already paid out in temporary total-disability. A worker who is temporarily disabled for four years in any of those states and has gotten the weekly maximum for temporary total-disability would get nothing for the loss of an arm.

Table 5
Rates and Benefits—1980

State	Employer Pays (*per $100 of payroll, selected manufacturing occupations*)	Injured Worker Gets (Maximums): *Temporary Total Permanent Total Death Benefits (per week)*	Injured Worker Gets (Maximums): *Loss of Arm to Shoulder*
Oklahoma	$3.63	$155.00	$22,500[a]
Texas	3.25	133.00[b]	26,600[a]
Louisiana	2.91	163.00	32,600[a]
California	2.56	175.00[c]	29,488
Vermont	.85	208.00	44,720
Wisconsin	.96	249.00[c]	35,000
Illinois	1.69	376.33[c]	88,438
Ohio[d]	1.69	275.00	30,938

[a] Amount paid out in temporary total-disability deducted from award for loss of part of the body in these states.
[b] Maximum total benefit in Texas is 401 weeks of compensation, or $53,333.
[c] Death-benefit maximum: California, $75,000; Wisconsin, $74,700; Illinois, $250,000.
[d] Ohio has an exclusive state fund.

Source: Rates—Confederation of State Manufacturing Associations Alexander Grant & Co. study.
Benefits—U.S. Chamber of Commerce

Nevertheless, raises in benefits are often used by insurers as an excuse to hike rates. When Illinois liberalized its benefits in 1975, the state's compensation insurers upped their rates by 46.8%. But, as the chairman of the state's Industrial Commission noted, "Not any one of the benefits has been increased 46.8%."*

The deputy director of the state's Department of Insurance at the time, Dave Taylor, was quoted in the Chicago *Sun Times* as saying that his department accepted the rate changes with no evidence of their being needed. The state simply relied on the recommendations of the National Council on Compensation Insurance.

*Chicago *Sun Times*, November 4, 1975, p. 58.

Two and a half percent underwriting profit sounds modest enough. Some years compensation insurers don't even make that much; sometimes the underwriting "profit" is negative. This does not mean that insurers lose money, though. Insurers invest the money they hold in their loss reserve and unearned premium reserve, and they make huge amounts of money on their investments. A recent report by Hill and Hunter, commissioned by the U.S. Department of Labor, shows that property and casualty insurers (workers' compensation is the second-largest line of this kind of insurance) hold $90 billion in investments, from which they earn $10 billion per year. * Continuing high interest rates as of mid-1982 make these figures extremely conservative.

In reviewing the history of ratemaking, the authors of the report show that before 1946, insurers were not allowed even the 2.5% profit margin; it was assumed that a reasonable return on stockholders' equity would be made simply through investment income. The middle and late 1940s was the worst period in the century for investment income, and the insurance industry asked state regulators for the first time to be allowed a profit margin to make up for the falling off in investment income. Since that time investment income has skyrocketed, but the 2.5% underwriting profit margin has become imbedded in insurance regulatory tradition. The U.S. Department of Labor report concludes that an insurance company would earn a return to book-equity yield (a reasonable rate of return to stockholders) equal to the Standard and Poors 400 index if the underwriting "profit" margin were −11%.

Amazingly, state insurance regulators do not take investment income into account in setting rates. With the exception of Massachusetts and Minnesota, insurance commissioners look only at underwriting profit in deciding how much insurers will be allowed to charge employers.

As we have seen, the underwriting profit in 1980 for stock companies was 3.4% of the earned premiums, while the margin for mutuals was 14.4%. The combined underwriting profit for that year

* Raymond Hill and Robert Hunter. "Workers' Compensation Insurance Ratemaking: Regulation of Profit Margins and Investment Income." In completion of contract number 41 USC252c3, U.S. Dept. of Labor (unpublished).

for stocks and mutuals was $829,676,000. This sum does not reflect the true return on equity of workers' compensation insurers, since it doesn't show a nickel of return on investment. If these insurers were required to operate at a negative underwriting margin of −11%, they would have collected approximately $2,300,000,000 less in premiums in 1980 than they did collect.

The most thoroughgoing alternative to private workers'-compensation insurance which yet preserves the basic structure of workers' compensation is the exclusive state fund. States with exclusive state funds are Ohio, Washington, West Virginia, Nevada, North Dakota, and Wyoming. Self-insurance is allowed in all of these states except North Dakota and Wyoming.

The Ohio State Fund uses its investment income to lower its rates when possible. Higher benefits required by inflation are offset by higher yields in significant parts of the investment market, also a result of inflation. There are dramatic results from these policies. In 1980, the Ohio State Fund collected $600 million in premiums from employers and paid out $500 million in benefits. Nevertheless, the State Fund managed to keep assets of $3.1 billion. Most importantly, in Ohio only the needs of those injured on the job and business have to be consulted in adapting the workers' compensation laws to the changing times. In states where private insurers rule the roost, it is their own interest that usually takes precedence over the less politically sophisticated employers and workers.

Since the insurance industry's needs don't have to be taken into account, Ohio has developed a compensation system that gives most of the money paid by employers to the intended beneficiaries of the system—injured workers. And while Ohio premiums are low compared to many other states, Ohio workers get more money back per capita than any other state. Social Security Administration figures show that in total dollar amount of workers' compensation benefits paid, Ohio is second in the nation, following only California, with its much larger population. *

There are other advantages to an exclusive state fund. Since employers must buy compensation insurance from the fund, insur-

* *Social Security Bulletin*, September 1981. Vol. 44, No. 9, p. 10.

ance agents are not needed and their commissions are eliminated. Statistics show that state funds contest fewer cases, so litigation is less prevalent in state fund states. State funds are more in keeping with the social-insurance rationale of workers' compensation, since unlike private insurance companies, they can be run primarily for the benefit of a state's workers.

Beyond Workers' Compensation

Exclusive state funds are more reasonable and cheaper than private insurers, but they cannot solve some problems that are built into the fabric of workers' compensation.

Occupational diseases are not being compensated more in exclusive state fund states than in other states. These claims create litigation in Ohio and Washington as well as other states. And there is no evidence that workers' compensation provides a significant incentive for employers to maintain safe and healthy workplaces.

Despite its shortcomings, workers' compensation in some form is needed. Even if injured workers were to regain their right to sue their employers for negligence, some type of insurance scheme would be needed to provide immediate benefits regardless of who was responsible. The no-fault concept of workers' compensation is necessary in our mechanized society and it should be continued. But workers' compensation is not enough—it cannot protect workers in the most essential way, by stopping accidents and diseases before they happen.

8

The Rape of OSHA

Workers are uniquely vulnerable both physically and legally to their employers' negligence and misconduct. The employer controls the workplace, and if he decides to ignore health and safety hazards he is not answerable to his employees for any injuries they may suffer. Workers' compensation allows the employer to insure the risk and immunizes him from liability for the employees' injuries and illnesses.

The inability of workers to control the hazards with which they work was expressed in an interview with Tony Mazzochi, former health and safety director for the Oil, Chemical and Atomic Workers Union: "The reduction of workers when it comes to industrial health and safety is tantamount to working on a plantation in the ante-bellum South. You give up all your rights when you walk through the plant gate. The employer can subject you to whatever he wants; you don't even have the right to find out from him what you've been exposed to."

Workers' inability to control on-the-job hazards is matched by their legal disability. A worker cannot sue the person who injured him if that person is his employer. But unlike minors and the insane, who are also legaly disabled but whose rights can be pursued by a guardian, workers simply have no right to go into court against employers for on-the-job injuries or occupational diseases.

The federal government has regulated health and safety hazards for consumers for many years. The Food and Drug Administration and the Consumer Safety Product Administration protect consumers, who nevertheless have the right to sue manufacturers of products that injure them. Workers won the right to federal protection through the passage of the Occupational Safety and Health Act in 1970. Unlike other regulatory schemes that supplement the individual's right to sue, however, OSHA is the only line of defense for workers, the only deterrent to injury and death on the job.

Occupational safety and health finally boiled its way to the top of the national agenda in the late 1960s because the percentage of disabling job injuries per hours worked was increasing at a disturbing rate; by 1970 it was 20% higher than in 1958. During the late 1960s, 14,500 workers were killed on the job each year, while 2.2 million, by the lowest estimate, were disabled each year due to industrial injury. *More Americans were killed on the job between 1966 and 1970 than died in Vietnam during the same period.*

The OSH Act was necessary because the states had not done the job of protecting workers. As a 1970 Senate report said: "Nor has state regulation proven sufficient to the need. No one has seriously disputed that only a relatively few states have modern laws relating to occupational health and safety and have devoted adequate resources to their administration, and enforcement . . . in sum, the chemical and physical hazards which characterize modern industry are not the problem of a single employer, a single industry, nor a single state jurisdiction. The spread of industry and the mobility of the workforce combine to make the health and safety of the worker truly a national concern."

The politicians were ready for a law protecting workers. Both parties were hoping to get blue-collar votes in the 1972 election. Although President Nixon and his congressional allies tried their best to weaken the substance of the act while keeping its form, they were eventually forced to agree to a compromise that produced a relatively strong bill.

Liberals had to compromise too. One of the most significant was a provision allowing states to take back the occupational health and safety function, although under federal supervision. The law said that the states could resume the function if they provided protection "at least as effective as" federal standards.

Two other provisions of the OSH Act are worth noting. The research function envisaged by the act was placed in the new National Institute for Occupational Safety and Health (NIOSH). This agency was a reborn form of the old Bureau of Occupational Safety and Health (BOSH) and was bureaucratically located far down in the hierarchy of the Department of Health, Education and Welfare, now the Department of Health and Human Services. Another temporary organization was created, the National Commission on State Workmen's Compensation Laws. This commission was created to study the problems of workers' compensation and report back to Congress on its findings. After delivering its report it was disbanded.

OSHA represented an attempt by Congress to provide uniform national standards for workplace health and safety and a mechanism to enforce those standards. Workplace conditions were to be improved by uniform federal regulation, but legal redress for workers injured on the job was left to archaic and inadequate state workers'-compensation laws. These laws could not ensure safe and healthy workplaces. Thus OSHA fell heir to the whole burden of improving health and safety conditions for America's workers.

OSHA was administered in its first years by Nixon appointees whose motivations were questionable. Instead of using the agency to tackle really serious health problems, the early OSHA bureaucrats promulgated thousands of pages of extremely detailed regulations concerning such things as the shape of toilet seats in employee rest rooms and the type of coat hooks that could be used in employee coatrooms. Really serious hazards, such as exposure to cancer-causing chemicals, were ignored. Workers were not being helped and employers were being harassed by this misguided use of governmental power.

OSHA was touched with the Watergate taint of political corruption so pervasive in Washington in the early 1970s. George Guenther, head of the agency during the 1972 presidential campaign, sent a secret memorandum to Under Secretary of Labor Laurence Silberman saying that the great potential of OSHA "as a sales point for fund raising" had not yet been fully realized. * In the

* *Crisis in the Workplace*, p. 229.

same communication, Guenther vowed not to set any "highly controversial" standards, such as the cotton-dust standard, in order not to offend business interests.

OSHA changed course dramatically under President Carter's administration. Headed by Eula Bingham, the agency became professionalized. Bingham brought in scientists and doctors to fill top positions. She rescinded thousands of useless regulations and set realistic priorities. Major hazards were targeted for regulation. A real attempt was made to set standards for exposure of workers in toxic substances. Programs to educate workers about health and safety problems were begun. OSHA became a tough enforcer of health and safety in the workplace, and industry began to take the federal government role seriously and to make changes to avoid being cited by the agency. An effective OSHA was a novelty and a challenge to industries affected by OSHA health and safety standards. Many industries chose to sue OSHA in federal court rather than comply with agency regulations.

The advent of the Reagan administration in 1981 with its ultra pro-business policies turned OSHA around once again. OSHA's swings in emphasis from pro-business to pro-worker and now back to pro-business reflects the larger political and economic conflict between business and labor, conservative and liberal. In Washington, as in the rest of the country, there has always been a tension between those who want to protect workers from injury and disease and those who want to protect employers from the necessity of spending money on health and safety.

OSHA's Mission

The Occupational Safety and Health Act was designed "to assure as far as possible every working man and woman in the Nation safe and healthful working conditions." It empowered the Labor Department to regulate the workplace so "insofar as practicable . . . no employee will suffer diminished health, functional capacity, or life expectancy as a result of his work experience."

The act places a "general duty" on every employer to "furnish to each of his employees employment and a place of employment which are free from recognized hazards that are causing or are likely to cause death or serious physical harm to his employees." Every employer is also obligated to comply with occupational health and safety standards promulgated pursuant to the OSH Act.

With the passage of the OSH Act, the federal government for the first time assumed the responsibility of protecting all the nation's workers from workplace hazards. Those laboring in the most hazardous occupation, miners, were given special protection with their own law, the Mine Safety and Health Act, which created a separate unit within the Labor Department to regulate mines.

Responsibility for enforcement of the OSH Act is lodged in the Occupational Safety and Health Administration, headed by an assistant secretary of labor. OSHA is supposed to enforce the "general duty" clause of the act as well as to promulgate and enforce standards for worker exposure to physical and chemical hazards. OSHA must conduct workplace inspections and issue citations for violation of standards. The law provides that advance notice of inspections may not be given to employers and that workers or their representatives must be allowed to accompany the inspector.

If an employer wishes to appeal a citation he or she must file a notice of appeal within fifteen days of receiving it to the Occupational Safety and Health Review Commission, whose three members are appointed by the President. Appeal may be taken from the Review Commission to the federal courts by either an aggrieved employer or OSHA.

OSHA conducts regular inspections of certain high-hazard industries, but will also inspect upon the complaint of a worker or union about a possible safety or health hazard. If the OSHA inspector finds a serious hazard, the largest fine he can propose amounts to only $1000. The inspector cannot close down the operation, no matter how dangerous it is, without first getting a restraining order from a U.S. district court. Some state inspectors, for example in California, do have the right to close down an operation that presents an imminent danger of death or serious bodily harm to workers.

OSHA must give an employer a "reasonable" period of time to abate the hazard. If the employer appeals to the Review Commission, the OSHA order is stayed until the commission acts. Although the employer does not have the right to a stay during an appeal to the federal Court of Appeals, that court may grant him a stay. If that happens, the OSHA order could be delayed for years and workers denied protection while lawyers haggle over arcane semantic questions involving the wording of the OSH Act.

Setting Standards

The primary function of OSHA is to promulgate and enforce health and safety standards for the workplace. * The OSH Act sets a lofty goal for federal standards, the law says, "in promulgating standards dealing with toxic materials or harmful physical agents . . . [OSHA] shall set the standard which most adequately assures to the extent feasible, on the basis of the best available evidence, that no employee will suffer material impairment of health or functional capacity even if such employee has regular exposure to the hazard dealt with by such standard for the period of his working life."

Setting standards for worker exposure to harmful substances costs employers money, sometimes large amounts of money. The cost to the cotton-textile industry of complying with OSHA's cotton-dust standard is more than $500 million. Standards that protect workers from hazards may require purchase of new equipment, such as sophisticated ventilation systems or expensive guarding on machinery. Standards may change patterns of work in whole industries, as happened with asbestos-goods manufacturing.

American business is not used to having to spend money in order to comply with government-imposed health and safety standards. It's no wonder that the response of many industries that were affected by OSHA standards was to take on board more lawyers and sue. From an economic point of view, this was entirely reasonable. The longer a standard is delayed, the longer the company gets to hold on to money that might otherwise have to be spent on complying with it.

From the point of view of a government official trying to protect the health and safety of the nation's workers, the standards are extremely hard to get adopted and to make stick. Joe Velasquez, director of the AFL-CIO's Workers' Institute for Health and Safety, and formerly an aide to Dr. Eula Bingham, director of OSHA in

* Nicholas Ashford, *Crisis in the Workplace.*

the Carter administration, said in an interview, "The standard-setting process is a bitch. You've got to put together every regulation looking over your shoulder for an industry suit. You have to make sure that it will sustain an ultimate review of the Supreme Court. The standard-setting process is being thrown over to lawyers and accountants."

Cotton Dust

The U.S. Department of Labor estimates that 559,000 workers are exposed to cotton dust and of this group 83,000 will develop byssinosis (brown lung). Betty Smith, the cotton-mill worker in Erwin, North Carolina, is not a number in this group; she's totally disabled as a result of byssinosis and is therefore a number in a group of 35,000—the number of people who have developed brown lung and are totally disabled from exposure to cotton dust. Byssinosis is one of the reasons the Occupational Safety and Health Act was passed in the first place. OSHA was created to set workplace standards that would prevent exposure to harmful substances such as raw cotton dust. It's too late for Betty and for her husband, who died of complications arising out of byssinosis, and for the cousins and parents who perished in the service of Burlington Mills without knowing that their jobs were using up their lungs. But it may not be too late for the young people of the cotton-mill towns. The federal government has come to their rescue with a law to make the employer clean up the plant—finally, after all these years. Or has it?

The Byzantine history of the cotton-dust standard typifies what the American worker is up against in getting workplace protection from the federal government. Protected by liberals, rejected by pro-business types, the cotton workers have had their hopes raised only to be dashed again, then raised once more. Shifting with the political winds, OSHA began to regulate exposure to cotton dust, then relaxed its efforts, then redoubled them, and now is retrenching. The story of the cotton-dust standard is a paradigm of the problems, both political and legal, of OSHA standard-setting.

Under pressure from organized labor, and aware of the high visibility of the byssinosis problem, OSHA began the process of establishing a standard of maximum permissible exposure to cotton dust in 1974. Notice of proposed rulemaking was published in late 1976. The following year OSHA received written comments and held public hearings. Testimony was taken from individuals and groups that would be affected by the standard. The printed record of testimony and "exhibits" totaled 105,000 pages. While the general public was largely unaware of the controversy, those affected by cotton dust and OSHA regulations were engaged in fierce political warfare. Predictably, the cotton industry decried the proposed regulations as too harshly restrictive while organized labor denounced them as too lax.

OSHA issued its final cotton-dust standard on June 23, 1978. It allowed more exposure than desired by labor, but considerably less than requested by industry. In order to achieve the reduction in cotton dust to which workers were to be exposed, OSHA required employers to install, at their own expense, a mix of engineering controls, including ventilation systems, and work practice controls, such as special floor-sweeping procedures. The cotton industry was given four years to comply with the rules, and in the meantime employers had to supply workers with respirators. Workers who could not wear respirators for health reasons were required by OSHA regulations to be transferred to another position with a low cotton-dust level, but without a reduction in wages or benefits. It looked like there was going to be a lot fewer Betty Smiths in Erwin.

But the textile industry was not content with a compromise regulation. Through its two main trade associations it sued OSHA to overturn the cotton-dust standard. Industry argued that the benefits of a standard should be proportionate to its costs and that OSHA ignored this consideration when it made the standard. The cost-benefit argument is a familiar tool of decision makers: a project should be undertaken only if its benefits will exceed its costs. Cost-benefit analysis is familiar and comfortable to managers of business because it provides a rational method for making choices. You determine the dollar amount of costs and measure it against the dollar amount of benefits. A project should be undertaken only if the dollar value of the benefits is greater than the dollar amount of the costs.

If cost-benefit analysis were followed by OSHA, presumably it would have to assign a dollar value to the life of a textile worker and measure that against the cost of making the workplace safe. The agency would have to decide how much Betty Smith and her kind were worth. Would it be as low as $500 or as high as $50,000 for each worker? Industry would argue for the lower figure, of course. Industry is used to the workers' compensation value of workers' lives. By this standard the lives of cotton-mill workers are cheap indeed.

Organized labor's views of the cost-benefit approach to workers' health and safety were well expressed by Joe Velasquez: "The cost-benefit argument is all bullshit. It's simply a means to slow down the standard-setting process. The health of the worker has got to be what OSHA looks at. When you talk about cost-benefit what you're really talking about is counting bodies, which is just unacceptable. In other words, if we propose a standard, industry will say, 'OK, you can have it—but first show us twenty or twenty-five thousand dead bodies.' That's an unacceptable way to approach the protection of workers in this country."

The legal arguments against cost-benefit analysis were based on the OSH Act itself. Labor argued that Congress purposely did not enact cost-benefit considerations into the Occupational Safety and Health Act. Cost-benefit analysis is inappropriate, they maintained, when nonmonetary values are at stake—like the lives and health of the nation's workers.

The cotton-dust case reached the Supreme Court in 1980, at the end of Dr. Bingham's tenure as head of OSHA. She and her assistants agreed with labor that cost-benefit analysis was inappropriate as the guiding principle in setting worker health and safety standards. By contrast, Ronald Reagan's regime promoted cost-benefit as the panacea for all social-benefit programs. By a lucky (for workers) twist of fate, oral argument in the Supreme Court on the cotton-dust cases happened to be scheduled for January 21, 1981, the morning after Reagan's inauguration. The lawyers who prepared the case for the government were holdovers from Bingham's OSHA. They were persuasive advocates for the relatively strong cotton-dust standard. Reagan's aides, exhausted no doubt from planning the extravagant parties that marked the new President's inaugural, were not fast enough to change the government's position to reflect their views.

But OSHA was marked for castration by the new administration.

It was just a matter of time. In late March 1981, OSHA announced that it was reconsidering its cotton-dust standard, so that it could apply cost-benefit principles to it. At the same time, Thorne Auchter, the new head of OSHA, recalled more than 100,000 pamphlets that had already been sent out to OSHA's regional offices for distribution to cotton workers. The pamphlets warned of the dangers of cotton dust and told workers of their rights under the Occupational Safety and Health Act. OSHA then went back to the Supreme Court and asked it not to decide the cotton-dust case. No one who followed the workings of the tradition-minded Court could ever recall the government asking it not to decide a case that had already been briefed and argued by the government. This unprecedented request, along with the notice of proposed rulemaking, appeared extremely arrogant. Reagan's Washington newcomers seemed to have forgotten or not to have known that the Supreme Court is a coequal branch of government, jealous of its prerogatives.

But the temptation to try to withdraw the cotton-dust case must have been too great for the new masters of OSHA. Cost-benefit analysis was slated to be the hallmark of the new administration—at least when the benefits might accrue to anyone other than big business and the defense industry. It would have been far better for the Reagan people to not have any Supreme Court decision on record than to have one that ruled out the keystone of the new approach.

In July 1981 the Supreme Court ruled that OSHA's cotton-dust standard, the one the Reagan people were trying to take back, was in accord with the wishes of Congress as expressed in the OSH Act. In a footnote to its decision, the Court said that it would "decline to adopt the suggestion of the Secretary of Labor" that it should not decide the case. The cotton-dust decision is a thoughtful reevaluation of some basic principles of the Occupational Safety and Health Act.

Justice William Brennan, speaking for the Court, said that industry argued that the costs of the health standard should bear a reasonable relationship to its benefits. If this principle were accepted, OSHA would have to show not only that the standard addresses a significant health risk, one that would lead to serious impairment of workers' health, but that the reduction in risk is "significant" in light of the costs of attaining that reduction. In Joe Velasquez's words, there would have to be enough dead bodies to make it worthwhile for industry to pay the bill for cleaning up the workplace.

Labor and OSHA under the Bingham leadership argued that the OSH Act requires OSHA to create standards that eliminate or reduce health and safety risks "to the extent such protection is technologically and economically feasible." Once a hazard is identified, you don't need to wait until workers are dying. The AFL-CIO argued that, under the OSH Act, OSHA had no choice but to adopt the most protective health or safety standard feasible.

The Supreme Court decided in favor of labor and Bingham's OSHA. It held that Congress itself had defined the basic relationship between costs and benefits by placing the "benefit" of worker health above all other considerations. The only limitation on OSHA's duty to set the most stringent standards is "feasibility." Congress did not intend for OSHA to set a standard that was technically or economically impossible to achieve. Any standard that struck a balance different from the one decided by Congress would be inconsistent with the law. Cost-benefit analysis is not required because feasibility analysis is required.

The Court held that OSHA was required to protect workers by setting its standards to the greatest extent feasible. "Feasible" did not mean that OSHA had to do a cost-benefit analysis, but only that it had to set standards that were capable of being met. Congress knew, the Court said, that the standards would cost industry substantial sums of money, but Congress chose to impose those costs when necessary to create a safe and healthy working environment. The legislative history of the bill shows that cost-benefit arguments had been put to Congress and rejected. Said Senator Yarborough, of Texas, one of the sponsors of the original bill (which was introduced by Senator Harrison Williams of New Jersey): "One may ask, too expensive for whom? Is it too expensive for the company who for lack of proper safety equipment loses the services of its skilled employees? Is it too expensive for the employee who loses his hand or leg or eyesight? Is it too expensive for the widow trying to raise her children on a meager allowance under workmen's compensation and Social Security? And what about the man—a good hardworking man—tied to a wheelchair or hospital bed for the rest of his life?

That is what we are dealing with when we talk about industrial safety. . . . We are talking about people's lives, not the indifference of some cost accountant."

Senator Eagleton commented in debate about the Occupational Safety and Health Act, "The costs that will be incurred by employers in meeting the standards of health and safety to be established under this bill are, in my view, reasonable and necessary costs of doing business." Justice Brennan, speaking for the Supreme Court in the 1981 cotton-dust decision, went on to say that, "Other members of Congress voiced similar views. Nowhere is there any indication that Congress contemplated a different balancing by OSHA of the benefits of worker health and safety against the costs of achieving them. Congress thought that the *financial* costs of health and safety problems in the workplace were as large or larger than the *financial* costs of eliminating these problems."

OSHA was set up to benefit workers. If it were to protect them to the extent that Congress had envisaged for it, the agency could not balance the lives and health of workers against the profit margins of employers. Whatever might have been the arguments before the cotton-dust case was decided, now that the Court has spoken, OSHA's duty to the workers of America is clear—they are entitled to nothing less than the greatest degree of protection feasible. For OSHA to attempt to do less would be a betrayal of its mission and of the nation's workers.

And what does OSHA say about the Supreme Court decision that upheld the previous administration's policies? Thorne Auchter, Assistant Secretary of Labor for OSHA, the Reagan administration's top occupational safety and health official, said to me in a tape-recorded interview in November 1981, "In the cotton-dust decision they said 'cost-benefit analysis is not required because cost-effectiveness analysis is.' And they went on to say that you must also have a determination of economic feasibility. They went on to define economic feasibility: that is, that an industry can maintain profitability and competitiveness."

Auchter's paraphrase of the Court's decision distorts the critical point. The Court said that cost-benefit analysis is not required because feasibility analysis is. There is a world of difference between cost-effectiveness and feasibility. The Court defined "feasibility" as "capable of being done." This has nothing to do with whether it is cost-effective. The law requires the maximum protection that is capable of being instituted. Congress did not mean to require OSHA to count the dead bodies of workers before it decided that a standard was cost-effective. Congress did not want OSHA to set standards that were impossible to attain, ones that would have effectively put whole industries out of business if they were enforced. These would clearly be unfeasible standards. But Auchter now says that he is *required* to set standards that do not hurt the profitability and competitiveness of the regulated industry. He is clearly trying to bring the outlawed cost-benefit analysis in through the back door by calling it "cost effectiveness."

His statement about cost-effectiveness was not a misstatement. He said it again and elaborated on it: "The Court said cost-benefit analysis is not required because cost-effectiveness analysis is. Once we set the level of employee protection, then we have to analyze it from a cost-effective standpoint as to what's the best way to maintain that level of protection. These processes take time and yet they are required by law now if our standard is in fact going to be upheld in court."

The implication of Auchter's statement is that the main consideration in deciding on the level of protection for workers is what's going to be the cheapest way to protect them. Obviously, it is much cheaper to buy respirators for cotton-mill workers than to install ventilation systems. It doesn't matter that respirators are uncomfortable, unhealthy to those who have even slight problems breathing, or potentially deadly to people with heart conditions. It doesn't matter that the respirators are often ineffective or that workers often don't wear them because they are so uncomfortable. If that's the cheapest way, then, according to Auchter's reasoning, the Supreme Court makes OSHA clamp respirators on the Betty Smiths of the Carolina cotton mills. To prepare for this approach, OSHA has begun the process of reevaluating its policy of favoring engineering controls to regulate exposure to hazardous substances and is now preparing to start promoting respirators.

An Internal Memorandum

Behind Auchter's statements is the private thinking of the inner circles of the Reagan Labor Department, which was expressed in an internal memorandum prepared by the department's chief lawyer, the Solicitor of Labor. This document, leaked by sources inside the department to a Washington legal newspaper, outlines options for OSHA in responding to the Supreme Court's recent decision in the cotton-dust case and another decision concerning the lead standard. Three options are presented to the OSHA policymakers.

"Option A" is to immediately begin new rulemaking for any industry where "there is thought to be a feasibility problem." "Feasibility problem" is a code term meaning "will cost money to comply with the standard." The memorandum says, "Cost-effectiveness of the standard (including relative protectiveness of respirators and engineering controls) could be considered." The good thing about this option, according to the Labor Department's lawyer, is that it is the fastest and most direct way of "rectifying feasibility problems" for regulated industries. In other words, this option is the quickest way of saving industry from the necessity of spending money to clean up the workplace. Another good thing about it is that it would "minimize the impression that the Supreme Court decisions seriously threaten the Administration's regulatory reform program."

There are several drawbacks to the option of just going ahead with new rulemaking. For one thing, there may not be enough evidence yet to show that the old standard is not feasible, since industry hasn't yet done anything to comply with it. Also, the announcement of new rulemaking "may appear to reflect resistance to the Supreme Court decisions." The Solicitor of Labor does not worry about the actuality of the resistance, but only about the harm to OSHA's "public image" and the impact on unions, which would be "enraged" if they felt "that their legal victory is being undermined."

"Option B" is to wait for six months or a year before starting the process of relaxing the standards. This option is desirable because it "will avoid the appearance of resistance to the Supreme Court's decisions." The bad thing is the possibility that some industries might not be rescued before they have to spend money to comply with the old standard.

"Option C" is the most cynical of all: "Do not modify the standard at all, but develop an enforcement policy which provides as much flexibility as possible. Numerous techniques, including setting appropriate abatement dates, granting temporary variances, and allowing claims of specific economic or technological infeasibility, are available to moderate enforcement of the standard."

The best thing about relying on loosening enforcement rather than standards, says the memorandum, is that "The relatively low profile of enforcement activities allows greater flexibility and avoids adverse public reaction." The only problem with this option is again PR. " [R]elying on flexible enforcement does not clearly show industry that OSHA is sensitive to their problems, as would a reconsideration of the standard, and does not clearly demonstrate regulatory reform initiative."

When the cotton-dust decision came down, it appeared for a brief moment that the young people of Erwin and all the other cotton-mill towns would have a chance to grow old without the curse of brown lung. *But the Court cannot enforce its own decision*. The law says that OSHA must protect workers to the greatest extent feasible. But the people running the agency that was created to help workers have been doing their best to circumvent the intent of the law they were meant to enforce and to undermine the decisions of the Supreme Court.

Assistant Secretary of Labor Auchter takes a philosophical view of standard-setting. It is, he says, not any more important than any other function of the agency he runs. And it takes time, lots of time. But that's all right, Auchter continues: "And not only do they [standards] take time, but once they're done they're still not done because we have to continue and go back and reevaluate those. We're going to be opening up, reopening the record in cotton dust and lead. And some people are going to scream about that, but the fact is we need to do it because those standards—complicated, very very complicated standards. A lot of those decisions were made without a particular data base."

Auchter's OSHA argued this last point in its plea to the Supreme Court to not decide the cotton-dust case, saying that Bingham's OSHA didn't have enough data to back up the cotton-dust standard. The Court rejected this argument in another footnote, saying, "It is difficult to imagine what else the agency could do . . . [to back up its standards with objective data]."

While Auchter and his cohorts are using the scarce and precious resources of OSHA to reopen standards that were adopted only two or three years before, American workers are being exposed to thousands of substances that have not yet been regulated, including many that cause cancer and birth defects. The cotton-mill workers will follow in Betty Smith's painful footsteps, dependent on the mill for a job and forced to breathe air that will destroy their lungs.

The cost-benefit or cost-effectiveness argument is not original to the twentieth century. There have always been those who thought it was too expensive to make the workplace safe and healthy. Coal-mine owners in nineteenth-century England testified to a Royal Commission looking into conditions in the mines that it was indeed true that women worked underground in tunnels often twenty-seven to thirty inches high, and children worked in tunnels sometimes no more than sixteen inches high. But nothing could be done about it, because "the expense [of creating healthier working conditions] would be more than twice over what the coals would be worth after they were got out."

Enforcement

OSHA must establish and enforce health and safety standards. The agency also has the duty of enforcing the "general duty" clause of the OSH Act. The law requires that every employer must "furnish to each of his employees employment and a place of employment which are free from recognized hazards that are causing or likely to cause death or serious physical harm to his employees." Enforcement is impossible without surprise inspection of the workplace. Congress recognized this and placed in the law the provision that OSHA must inspect workplaces without giving advance notice to the employer.

But OSHA under the Reagan administration is reluctant to enforce the law. The agency is being geared down so that its actions will be agreeable to business to the maximum extent feasible. The current OSHA chief, Thorne Auchter says, "OSHA today is a cooperative regulator. We're not the enemy of business, we're not the enemy of labor. We're the enemy of safety and health hazards in the workplace. We're not interested in crime and punishment, we're interested in safety and health programs."

According to the AFL-CIO's Joe Velasquez, there are certain code words used by those who wish to make OSHA ineffectual. "Cooperative" is one of them. It sounds attractive. It means "no enforcement." Valasquez's view of OSHA's function is different from Auchter's.

> I think that OSHA has got to be a tool for working men and women to bring about workplace change. Obviously a cooperative approach to problem solving is most effective. But for it to work, you've got to have two parties who are equal in power. Workers have to have some kind of power to get the companies to listen and respond and the company has to have some kind of power to get the workers to listen to their side of the issue. When one side has more power than the other, then there can be no cooperation, no meeting of the minds. Employers by and large are far more powerful than workers. OSHA should provide a little bit of equalization of power. It should provide a resource for workers to get things done in the workplace.

The new OSHA doesn't see itself as a resource for workers or indeed as an enforcer of workers' rights. The agency has instituted a new strategy for enforcement that will effectively end OSHA's role as regulator of the safety and health of America's workers.

A Paper Tiger

OSHA was created with some formidable powers to make and enforce health and safety rules. The Nixon administration and most conservatives wanted a much weaker organization, but they lost the legislative battle and had to settle for a genuinely effective agency. Or at least potentially effective, because even the most powerful agency can be turned into a paper tiger if those running it are determined to destroy its effectiveness. The process of weakening

OSHA is two-pronged. The first prong is to not set any more health or safety standards and to recall those that have already been set. The second is to eliminate enforcement of standards and employers' general duty to the maximum extent feasible.

In late July 1981, OSHA held a high-level meeting in Colorado Springs to unveil the new order to its field staff. A new procedural manual was passed around and the staff given their marching orders. The new "cooperative" OSHA was to be less "adversarial" and much more "understanding" of industry's problems. The new spirit was aptly captured by a directive in the manual: "Assessment and collection of penalties has resulted in antagonism among employers and the general public and has been an impediment to the successful attainment of the goals as set by Congress. Therefore it is considered appropriate to revise the penalty procedure." The working details of this delicate sentiment, spelled out in the manual, are that employers who have been cited for health and/or safety hazards will not be fined if they have abated the hazard, or even if they merely say that they have.

Employers who are cited will be pretty rare anyway under Reagan's OSHA. General-schedule inspections, the routine inspection of manufacturing establishments for safety hazards, have been severely curtailed. Only those plants in "high-hazard" industry will be subject to these inspections. All "low-hazard" industry has been put on notice that it is exempt from routine inspections. Most plants in high-hazard industry will be exempt also.

Which industries and which plants within an industry will be inspected depends on "lost workdays." All industries have been reviewed, and those that have a lost-workday rate below 4.2 days per 100 workers per year are declared low-hazard. In high-hazard industries, the inspector will begin his or her inspection by first looking at the employer's OSHA log, a record of job-related injuries and illnesses that employers who have more than ten employees are required to keep. If the log shows a lost-workday rate below 5.2 per 100 workers, the national average for manufacturing, the inspection is ended before it is begun. The inspector simply closes the books and goes home. However, even establishments that have a lost-workday rate of *more* than 5.7 per 100 workers are not inspected unless there is a complaint.

According to Bureau of Labor Statistics estimates, at least

279,400 manufacturing employers, 86% of all manufacturing firms in the United States, would be exempt from regular OSHA inspection, either because they were in "low-hazard" industries or because the individual employers had an average or better lost-workday record.

OSHA doesn't bother to use any criteria other than lost workdays in determining whether to conduct a safety inspection in a high-hazard workplace. But lost workdays do not include on-the-job *deaths.* A plant that has had several fatal accidents but not over 5.7 per 100 lost workdays would be automatically exempt from inspection. Just as in workers' compensation, it pays the employer to kill a worker rather than just injure him or her, since a dead worker will not be held against the employer, while one in the hospital will be accumulating lost workdays.

In computing lost workdays, the new OSHA aggregates all of a company's employees. White-collar workers are lumped together with blue-collar workers. In the Pittsburgh area, where many steel companies have their headquarters, this might result in steel mills being exempt from inspection because there were so few lost workdays among a steel company's office workers.

Lost workdays as the criterion for inspection suffers from another major fault. The information on which OSHA bases its decision of whether to inspect is furnished entirely by the employer. Marginal businesses will have a very strong incentive to falsify their records. Some employers will call in their injured workers and make them report to work. Workers with broken arms or legs will be told that they must show up or they will lose their jobs. They would get their regular pay but not be required to do anything, just punch the time clock in the morning and at the end of the workday, and stay on the premises of the plant in the meantime. This has been done for years by some employers to keep their workers' compensation rates down. Dr. Molly Coye, a National Institute for Occupational Safety and Health (NIOSH) scientist, recalls visiting a plant that had a big sign over the front gate: 178 DAYS—NO ACCIDENTS. In touring the plant she came upon a room filled with men wearing casts sitting around. These were fresh injuries, but not "lost workdays," and not counted as work accidents.

OSHA checks the employer's log of injuries against workers' compensation records. This is not an easy bureaucratic task, since each state has its own unique system for reporting workers' compensation claims. Some do not even require such reports. In these states there will be no corroboration of the OSHA log.

There is another serious problem with the lost-workday measure as the trigger for an OSHA inspection. Under workers' compensation, an off-the-job injury does not result in lost workdays for purposes of experience rating. This has encouraged unscrupulous employers to contest compensation claims by asserting that the worker's injuries were not job-related. It is up to the worker to prove otherwise.

The story of Dorothy Hanna, the steelworker in Indiana, illustrates how some employers try to juggle lost-workday numbers. After Dorothy had been out of work for almost a year due to the injuries she sustained when a large piece of metal fell on her, she was summoned to the company doctor. He told her that she was now able to go back to work. With a fine disregard for logic, but a sharp eye out for lost-workday statistics, he also said that she could collect disability payments under the company's non-work-related injury-compensation program. Still disabled from her work accident, Dorothy resisted and had to litigate her case. If she had chosen to take the alternative program, as most workers would, since they need the money, her further time off from work would not be considered "lost workdays" for either workers' compensation purposes or for OSHA. Her employer was Bethlehem Steel.

The employers who are inclined to cheat on their records are informed in advance by the new OSHA on how to do it. OSHA's form letter to employers in high-hazard industries says that:

> if the following conditions are met, your establishment will be exempt from a scheduled walk around inspection on this occasion:
>
> (1) Your workers' compensation records indicate that your company's injury rate is below the national manufacturing lost work day case rate (5.2 lost work day cases per 100 workers) and,
>
> (2) There is general agreement between your establishment's workers' compensation injury records and the corresponding OSHA-200 log data.

One more advantage is offered to the employer whose records look good. "Finally, if these conditions are met, but a recent complaint has been filed, the compliance officer [inspector] will inspect only those items noted in the complaint."

All record keeping is in the hands of the employer: OSHA has no capability to supervise it. If a company falsifies its records, OSHA will have no way of finding out about it.

Finally, the statistics that determine which industries are "low-hazard" and which "high-hazard" are compiled by the Bureau of Labor Statistics, whose budget is being cut. Therefore, the reliability of the BLS survey upon which the statistics are based is more dubious than ever.

"Targeting" OSHA enforcement appears at first glance to make sense. The agency cannot inspect every workplace in America; it is chronically short of inspectors and probably always will be. So why not inspect the worst? But it is just because of OSHA's limited resources that the targeting policy is a disaster. Its greatest value in terms of enforcement has been as a deterrent—a potential threat to employers who might be tempted to allow serious hazards to continue to exist in their workplaces. OSHA is also a rationale for responsible management of large corporations to sell health and safety to possibly reluctant stockholders. The deterrent effect OSHA inspections used to have was like that of an IRS audit. Few taxpayers are actually audited, but most are kept honest by the fear of one. If the vast majority of taxpayers were exempt from audits . . . but the idea is unthinkable because the government is genuinely committed to keeping the IRS effective.

Until the vast majority of manufacturers became exempt from safety inspections in Octocber of 1981, OSHA's influence was significant. Allan Tebb, director of the California Workers Compensation Institute, confirmed this, saying that employers became more safety conscious than ever before. They call in insurance company loss-control experts to advise them on how to clean up their workplaces much more than ever before. Employers were afraid of being cited by OSHA. Most can now relax in the certainty that they are exempt.

Behind the details of changes in policy, the loosening of health and safety standards, the relaxation of enforcement, is a pervasive lack of knowledge in OSHA's top management of the technical background of the occupational health and safety field. Dr. Eula Bingham, OSHA's former director, a scientist and expert in the field, sees this as a deliberate strategy on the part of the Reagan administration. She told me in early 1982:

> The net result will be that OSHA will fall into the kind of disrepute that it was back in the early 70s, only more so. It will be ignored and it will have no meaning. But I think that's part of the purpose. I think that it was by design that trained individuals, even from a corporation, were not brought in to run OSHA. There were Republicans who were from industry or other groups that were trained in occupational safety and health that could have been brought in to run that agency and to staff the major positions. But there is only so far a qualified professional will go. I know half a dozen qualified corporate medical directors who will agree that a certain standard is necessary. We may disagree on the level, but it won't be by a very large factor. There is a body of professional knowledge that understands what needs to be done. I guess that the administration knew that and that wasn't what they wanted in there. They wanted to really gut the agency. Personally I think they've been very successful. There has been a brain drain in terms of science to other agencies and to industry. I guess they wanted that to happen. The scientists that stayed have been intimidated.

While it is sporting of the President to give the job of head of the Occupational Safety and Health Administration to an amateur, the working people of the country will have to suffer the consequences of his having to learn on the job. Thorne Auchter himself admits that he never had any contact with OSHA on the Washington level before he moved into the assistant secretary's suite. His only experience has been helping to run his family's construction business in Florida and, most importantly, as special events coordinator of Ronald Reagan's Florida campaign.

Auchter prides himself on his management skills; his spacious office is decorated with flow charts and graphs. But he doesn't seem to have a point of view, a reason to manage. In private industry it's simple. The purpose of it all is to make money, and you can usually see the results of your work, up or down. But government regulation isn't like that. It requires a firm grasp of priorities and the ability to choose paths of action from a multitude of choices. People who do not have that sense of priorities, who do not know why they're doing what they're doing, concentrate on process. What interests them is the mechanics of the operation, the sequence of papers flowing in and out of in and out boxes. So it is with Auchter. When asked what his objectives for OSHA were to be in 1982, he had a hard time responding:

> Well, they're various and sundry things that we're going to be looking at and it's all based not on a concrete objective. What I'm looking for is trends. I want to see, for instance, continued reduction in the rate of contested cases. I've got various and sundry deadlines for some of our program directors on things that I want them to accomplish. Some of them are supposed to come back to me with suggested particular emphasis programs that I've asked them to work on, Federal agency programs, operations for instance, and some other things. We're going to be doing some new things with our training institutes and I've got some deadlines for those people that are involved with that to come back to me with option papers on various parts of the way we train people and new management techniques that we're going to make available to our people at our training institutes.

Nevertheless, Auchter is proud of his work. His boss, Secretary of Labor Raymond Donovan, is an amateur also and the former owner of a construction company. He is not likely to criticize Auchter for failings that he shares, like lack of depth of knowledge of occupational health and safety or of the way government works. Donovan, who never worked for government before, took over an operation that had a budget of over $23 billion. As Auchter says, "We wiped out the whole senior management of the Labor Department. Whooosh, it was gone with the election on November 4. And we came in, the new senior managers, and in less than a year we have changed the direction of this tremendous ship [the Labor Department], if you will. That requires a lot of energy and we've put in a lot of long and detailed and decisive hours here."

But with all the hours he's put in, Auchter doesn't know some basic information that you would think the nation's number-one occupational health and safety official would know, or at least have an opinion about. When asked what he felt were the most serious hazards facing American workers today, Auchter responded, "*I have not looked at that at all*, I may look at it some time in the future, but it will be based on some data. It won't be based on some esoteric subjective judgment. And if I had an answer for that I would be very careful about how I phrased it. . . . Regulatory agencies have spoken in the past about suspicions or suggestions or areas that they're getting ready to look into and all of a sudden sales of companies involved in that operation drop off . . ."

Government officials do not have to be ignorant. Witness Dr. Bingham's answer to the question of what are the most serious hazards facing American workers:

> There [are] a number of serious things facing American workers: toxic chemicals pose quite a serious hazard whether the condition is cancer or reproductive problems in men and women, or perhaps some of the chronic diseases like chronic kidney disease or chronic pulmonary disease are among the most serious occupationally induced conditions.
>
> After that I think that we have to take a hard look at some of the conditions that occur as a result of the human body structure not having been taken into account in the design of machinery. You know, people bend to fit the machine rather than the machine being designed to fit the workers. As a result we have carpal tunnel syndrome, back injury, and a whole variety of chronic injuries that significantly affect the health of workers.

> Trauma continues to be a major problem in this country, even though there have been some inroads. As we try to force production higher, for example as we keep running refineries day after day after day with little or no maintenance, as a society we become very suicidal because valves deteriorate, the equipment is in poor condition, so we run the risk of toxic exposure where pressure is built up. Electrocutions remain a problem.

Lack of expertise on the part of an agency head is a serious failing, but not necessarily fatal. Secretary of Labor Donovan could have made up for it by filling the top department job with experts. He chose people with no experience in government, like Thorne Auchter. Auchter could have made up for it by appointing experts to the top jobs in OSHA. Instead, he chose amateurs like himself. The result of such insouciance is an institutional inability to respond to the mandate of the agency. Perhaps Eula Bingham is right, and it is by design that the Labor Department, which is meant to represent the interests of the nation's workers, has been put in the hands of people who don't even know what the issues are.

In 1980, under the Carter administration, the Labor Department published a report to Congress that said that there were 100,000 annual deaths in the United States caused by occupational disease. The report has been widely read and criticized. Some say that the figure is too high, some that it is much too low. When asked his opinion of this report, Auchter said in late 1981: "I have not read the report. It has not been a topic of discussion. As a matter of fact, the only time I've seen it brought up was in an editorial in a Mississippi newspaper that just went across my desk this week as part of our review of news clippings. That's the only time I've seen the question raised. Nobody in Congress or labor or management has ever asked me to take a look at that particular subject."

"That particular subject" is not an abstraction. It is the shocking reality of death on the scale of a major war. It is death, not by gunfire but by disease, not because of malice but because of institutionalized greed. And in this war the workers' generals are working for the enemy.

9

States' Rights and Wrongs

The issue of state versus federal power is an enduring theme in American political life, a persistent idea inherited from an earlier and simpler era. But the problems of the latter part of the twentieth century are so complex and difficult that they can be effectively addressed by the public only at the federal level. Leaving worker health and safety or workers' compensation to state control really means continuing the status quo, for the states do not have the resources, either political or technical, to make significant changes. Business and the insurance industry, themselves organized at the national and international level, favor state "control" because it cannot challenge their own real control.

If organized labor and the workers it represents, both organized and unorganized, have any political clout, it is at the national level. The AFL-CIO and some of the large individual unions maintain lobbyists and staffs of experts whose job it is to follow and try to influence federal legislation. Few state federations of labor are able to maintain an effective lobbying effort in their state legislatures. This weakness is particularly damaging in those states where labor is poorly organized, and where the problems of working people are most pressing—in those states most in need of an advocate for workers.

The important questions of occupational health and safety and workers' compensation are not technical or ideological, but political. It all boils down to the issue of who is in control. Control of the workplace is the key to all elements of the problem of worker health and safety. Control of the legal and financial mechanism of compensation is the issue of workers' compensation.

The political battle for OSHA was fought between the forces of business and organized labor. The act finally wrested some degree of control of the workplace from business and gave it to the federal government's Occupational Safety and Health Administration, whose mandate is worker protection. Under the act, employers were given the responsibility of providing a relatively safe and healthy workplace. For the first time, employers were required to obey federal standards of worker exposure to certain identified hazards. Workers and unions were given the right to invoke OSHA's enforcement mechanism. The employers' monopoly of power in the workplace was ended.

The workers who won a victory with the passage of the OSH Act were up against a more formidable political foe when it came to workers' compensation. While big manufacturers and the chambers of commerce fought OSHA, it was the giant insurance cartel, smooth and united, that defeated all attempts to alter the power relations in the workers' compensation arena. Also, with OSHA, Congress had acted to fill a vacuum. But with workers' compensation, Congress faced a system that was already entrenched at the state level. Federalization of workers' compensation had to come to terms with the incredible complexity of the system, layered as it is with seventy years of the laws and practices of fifty states. Workers' compensation reform on the federal level has been postponed indefinitely.

The issue of state versus federal control is a glaring anachronism in light of new technology, which is bringing with it ever more complex dangers to the workplace. American industry, with its instinctive rush to embrace the new and untried, is imitating the sorcerer's apprentice. Able to summon magical powers, industry cannot stop them. What was a blessing has become a menace. Even future generations, the unborn and yet unconceived, are now at risk because of industry's actions—and government's inaction.

Reproductive hazards represent the new frontier in occupational disease. Dr. Peter Infante, an OSHA scientist, said at a conference on occupational disease in 1979, "There is substantial evidence that certain agents found in the occupational setting can affect normal sexual functions and the ability to produce healthy offspring."* Since the effects of Thalidomide were understood in the early 1960s, the science of teratology, the study of birth defects, has become of major importance. The danger of Thalidomide, which caused children to be born with atrophied arms or legs, sometimes with no limbs at all, alerted scientists to the potentially hideous effects of new chemicals. The new science of teratology has begun to raise alarms about reproductive hazards in the workplace. Some companies, recognizing their potential liability to lawsuits, have banned women of childbearing potential from workplaces where they might be exposed to teratogens.

Besides birth defects, substances found in some workplaces have the power to change genetic material in ways that are still not well understood, to destroy or damage sperm, to cause infertility in women and sterility in men, to cause growth retardation in fetuses, and also in those children born with an occupational disease. Even behavioral changes may be detected in children born with exposure to some substances. Some defects are not apparent at birth. Only when the child is three or four do the parents find out that their little girl or boy has a congenital heart defect, for example. One of the most common results of exposure to a toxic substance of a pregnant woman is spontaneous abortion. In many cases, the woman does not even realize that she has been pregnant.

Concern for the problem of reproductive hazards developed later than concern for cancer. Therefore, the ability of medical science to measure carcinogenic effects is considerably more advanced than the ability to measure reproductive hazards. The number of sub-

*"Lost in the Workplace: Is There an Occupational-Disease Epidemic?" Proceedings of a seminar for the news media, September 13-14, 1979. U.S. Dept. of Labor, Occupational Safety and Health Administration.

stances that have been tested for reproductive effects is very small. Most of the data are based on animal tests.

The responses of industry and government to the danger of reproductive hazards in the workplace illuminate the hollowness of the choice between federal and state control.

Industry's most forceful response to the problem was to fire women of childbearing potential unless they were willing to be sterilized. For example, the Bunker Hill Company of Kellogg, Idaho, and American Cyanamid, at its plant in Willow Island, West Virginia, adopted policies that require women who work in certain areas of their operations to undergo sterilization or lose their jobs. Some women have chosen to be sterilized. The companies argue that lead and other substances in their plants have been proven to cause defects in fetuses. The only reliable way of preventing the birth of deformed babies, they say, is to remove women who are potential childbearers from the workplace.

The movement to exclude women from many workplaces began to gather momentum in the middle 1970s, but was checked by Eula Bingham's OSHA, backed by organized labor. Exclusion of women solved nothing, since many of the substances that deform a fetus *in utero* also alter sperm and cause sterility in men as well as women. The exclusion of women of childbearing potential didn't protect the offspring of male or female workers who remained on the job, and it discriminated against women who needed jobs. Interestingly, the movement to exclude women applied only to women working in traditionally male jobs. Women in traditionally female jobs were nowhere excluded—for example, nurses exposed to anesthetic gases in operating rooms.

The only governmental response to the problem of reproductive hazards in the workplace came from OSHA and from NIOSH, the National Institute for Occupational Safety and Health—from the federal government. It is not surprising that no state responded. First, the problem is relatively new in most industries and the data to back up government regulation are scanty. No state has the scientific resources to back up any regulatory actions, nor will any state, except perhaps California, ever have the resources to challenge industry on the technical aspects of the toxic effects of thousands of substances. It is absurd even to speculate further on this line. The research capability to establish the danger of workplace substances should not be duplicated, even if it could conceivably be duplicated in state after state.

To return OSHA's functions to the states would mean that the problem of reproductive hazards in the workplace will not be dealt with. Before OSHA, worker health and safety problems were simply ignored by government—precisely because there was no federal involvement. If OSHA's functions were returned to the states, the same situation would quickly assert itself, and workers and the unborn would be defenseless.

Workers' compensation, still subject to the vagaries of state legislatures, simply has no ability to respond to the danger of reproductive hazards in the workplace. Historically, it pays only for inability to work, not for any noneconomic loss. Loss of sexual function or loss of ability to produce healthy children is not conpensable under workers' compensation, except for some rather rare provisions such as Indiana's anomalous award of $3700 dollars for the loss of one testicle or $11,250 for the loss of both. Nevertheless, workers' compensation is the "exclusive remedy," barring all lawsuits by workers against their employers, no matter how negligent the employer is in exposing his workers to reproductive hazards.

Children of exposed workers are not covered by workers' compensation, however, since they are not employees. Therefore, they are legally capable of suing the employer for negligence if they have been damaged by the exposure of their parents to toxic substances. Although such suits are rare, they have been brought and won by children with birth defects resulting from a parent's exposure at work.

The policy of excluding women of childbearing potential from the workplace thus begins to make sense. Employers like Bunker Hill and American Cyanamid know they are immune from suits by their employees. And they know that the state of science is not advanced enough for a plaintiff to be able to establish to the satisfaction of a potential jury the causal link between a hazard at work and the defect of the child of a male worker. But the company might well get nailed by the child of a female worker, a child who as a fetus was damaged through the mother's exposure to workplace pollution.

One of the original goals of workers' compensation was to internalize the costs of occupational injury and disease and thus encourage employers to maintain a safe workplace. However this may have worked in the early part of the century, there is little doubt that it doesn't work today. That it costs employers absolutely nothing to damage or destroy the reproductive potential of their own workers outlines in bold relief the inadequacy of workers' compensation to help make the workplace safe. It is a shield for employers and it offers nothing to a worker whose sex life or reproductive capacity has been damaged at work.

It is hard to imagine any state changing its workers' compensation laws to make reproductive damage compensable. The coalition of manufacturers and insurance companies that would be sure to oppose such a change in the state legislature would be politically unbeatable. The federal government, on the other hand, does have the institutional ability through OSHA to regulate exposure to reproductive hazards, given the political will of the administration that is running the agency. Although Reagan's OSHA will not tackle this problem, a future administration may. Like it or not, it is apparent that only the federal government, not the states, has the capability to respond to the problems of the late twentieth century.

The National Commission

The framers of the Occupational Safety and Health Act knew that something had to be done about workers' compensation. But since little was known about how it was working in the states, they decided to create a commission to study the problem. The commission issued its final report in July of 1972 and disbanded soon after. The National Commission on State Workmen's Compensation Laws created the ideological framework for later proposals for reform of workers' compensation and therefore cannot be ignored.

The commission was heavily weighted with conservatives and insurance industry representatives. Only two out of twenty members represented labor. Traveling around the country holding public hearings, its members discovered that workers' compensation was a shambles. And they found that the question of federal versus state control of workers' compensation was central. As Michael Peevey, former AFL-CIO official and member of the commission, says, "Whether to federalize workers' compensation was the key issue before the commission."

Bowing to the weight of the evidence, the commission found that "The inescapable conclusion is that State workmen's compensation laws in general are inadequate and inequitable." Neverthelesss, it favored retention of state control. The commissioners rationalized this position by saying that they believed in the basic principles of workers' compensation but were disappointed that "present practice falls so far short of the basic principles, and because there is no possible justification for this shortfall."

Consistent with its temporizing in other areas, the National Commission decided not to make any recommendations about the issue of permanent partial-disability (PPD). PPD's gross inadequacy and the litigation this category of benefits engendered created the widespread discontent with workers' compensation that led to the creation of the National Commission, yet it was precisely this issue that the commission chose to ignore. There should be *another* commission, the commission recommended, and *that* commission should decide what should be done with PPD.

Recognizing that the states might not choose to reform their own workers' compensation systems, even to the mild extent it suggested, the commission recommended that they be given three years to do this. Congress should mandate federal standards for state workers' compensation laws by 1975 if the states didn't clean up their act. The National Commission did not advocate federalization of workers' compensation, but a more modest role for the federal government as guarantor of minimum standards, a similar function to the one the federal government performs with unemployment insurance. The two members of the commission from organized labor disagreed with the 1975 date. They thought that federal standards should be enacted right away—there was no need to wait another three years. One important result of the National Commission's work is that a number of states, afraid of national standards, raised their temporary total-disability payments. Bobby Barnes, former labor union official and now a workers' compensation commissioner in Texas, told me that his state began to reform its workers' compensation system because legislators and the insurance lobby feared the imposition of federal standards.

Following the National Commission's recommendations, bills were introduced in Congress to set federal standards for workers' compensation. At first the insurance industry paid lip service to the idea, but it soon dropped the façade of reasonableness. The federal-standards bills became weaker and weaker, but by the time the third and weakest bill was introduced in the Senate there could be no doubt what the position of the insurance industry was. When informed of the total opposition of the American Insurance Association, the largest insurance-industry lobbying group, Senator Jacob Javits, one of the two senators most influential in the reform movement, complained that he had drafted a new bill that met all the important objections the organization had raised to previous bills. "In fact," Javits said, "a great deal of this legislation reflects the comments and suggestions AIA has made in the past." The bill was so sensitive to insurance interests that it was opposed by the AFL-CIO. But the insurance industry could not be placated; it simply did not want any federal involvement in workers' compensation.

The Carter administration finally came out in favor of federal standards for workers' compensation. With its demise, and the political demise of the two major congressional champions of federal workers' compensation standards, Senators Jacob Javits (defeated for reelection) and Senator Harrison Williams (disgraced by ABSCAM) and the advent of the antilabor Reagan administration, the movement for federal standards came to a dead stop.

Regression

Regulation of the workplace was made a federal responsibility by the Occupational Safety and Health Act in 1970. But there was a compromise: states were allowed to take over the job of insuring worker health and safety under close federal supervision. If after a few years a state was found to be able to establish a fully effective program of regulation, the state would be allowed to implement its own program and federal OSHA would withdraw. Organized labor opposed the concept of state enforcement, arguing that the main reason for the OSH Act which brought the federal government into the business of regulating workplace health and safety was the miserable record of the states in this area. The states could not be trusted to protect their workers. A hundred years of experience proved their inability or unwillingness to do the job.

After the Occupational Safety and Health Act became law, twenty-four states set up their own job health and safety operations. A large part of their budget came from the federal government. Most of the large industrial states, where organized labor is strong, chose not to develop their own programs but to allow federal OSHA to safeguard workers on the job. Indiana is an example of a state that chose to develop its own program. A conservative rural state with an urbanized blue-collar northern fringe, Indiana was not a good bet to have an effective program. Its workers' compensation system is inefficient and has among the lowest benefit levels in the country.

Indiana's record of concern for the safety of its workers does not inspire their confidence. Larry Keller, an official of the United Steelworkers, helped organize a worker safety and health conference for the state AFL-CIO in late 1980. Indiana's chief factory inspector attended with seven of his staff. One of the union people asked the chief inspector what he would do if a local union requested a copy of the state's inspection report of a violation of health or safety rules. Since the company got a copy of the report, could the union also obtain one? The Indiana state official replied that no, the union could not get a copy of the report—the state could not afford the extra postage.

Not surprisingly, union officials and workers nurture strong suspicions of the Indiana program's good faith as well as its lack of effectiveness. Workers at one plant saw an inspector come for a scheduled inspection but not enter the plant. The inspector stopped at the gate and waited for a few minutes. The personnel director came out and the two went to lunch together. The inspection never took place. In a steel mill, workers in the foundry were breathing vapors dissolving from chemical coolants. They were getting sick and were worried because the chemicals had only recently been introduced and no one knew what they were. The workers asked the state to send an industrial hygienist. Instead, the state sent a clerical employee who knew less than the workers themselves. In another case an inspector who was really just a clerk was sent to inspect hazards at a cement factory.

It's no wonder that Indiana's attempt to take over the occupational safety and health function ran into opposition. In 1977 the state and local AFL-CIO and the United Steelworkers petitioned OSHA to withdraw Indiana's state plan. In 1980, the agency, then headed by Eula Bingham, began the process of withdrawing the state's authority in this field.

There was, OSHA said, a consistent pattern of poor performance in important program areas. OSHA charged that the state of Indiana:

1. Failed to recognize large numbers of serious hazards in workplaces that it inspected.
2. Imposed substantially lower penalties for violation of the law than federal penalties for the same violation.
3. Failed to hire enough experienced, qualified industrial hygienists, which resulted in inadequate health enforcement.
4. Failed to fully protect workers' rights. *

All of Indiana's shortcomings came to light while there was still a federal presence in the state's workplaces. Until final approval of a state's program, federal OSHA is supposed to monitor the state's performance to make sure that it is adequate. Indiana contested OSHA's charges and insisted on litigating the issues. The state's program was saved not by the courts but by the Reagan administration, which dropped the charges against Indiana and allowed the state to continue to run the program, without even a promise by the state to do better in the future. Reagan's OSHA went further—it withdrew the federal OSHA presence.

The workers of Indiana, particularly in the state's dangerous steel industry, are worried. They remember the bad old days before there was an OSHA to help them, and they are afraid those days have returned.

With the federal agency available, workers could get an inspector within twenty-four hours if there was a life-threatening danger, and they could get an inspector who knew what had to be done. Now that the federal inspection machinery has been dismantled, there is

* *Occupational Health and Safety Letter*, January 22, 1981, p. 7.

no one to make sure that Indiana lives up to its responsibilities to its workers.

The Politics of Workers' Compensation

What happens to workers' interests when the scene of action is the fifty state capitols is illustrated by the politics of workers' compensation. While details vary from state to state, the cast of characters is the same and the relative power of the interest groups is similar in most states.

It is noteworthy that any changes in workers' compensation come about solely at the instance of interest groups. Few state legislators have any understanding of the simplest notions of workers' compensation. Unlike Congress, they have tiny staffs and only the sketchiest of technical resources. Legislators simply have to get their information from lobbyists, who are usually richer, smarter, and more influential than the legislators themselves.

Each of the fifty states has its own constellation of economic and political centers of power, but where there is no exclusive state fund a common pattern of groups comes into play when workers' compensation becomes an issue. They are: the insurance industry, the general business community, including large manufacturers and the smaller businesses usually represented by the Chamber of Commerce; and organized labor. In many states the trial lawyers' association plays a major lobbying role, as well as medical societies whose member doctors treat or examine workers hurt on the job. Usually each group lobbies only for the narrowest and most selfish interest of its constituents.

The impetus for change in workers' compensation laws usually comes from the insurance industry. This great economic force has the most direct interest in workers' compensation and it is highly organized. The insurers and their lobbying groups employ the best and the brightest in every state. Their expertise usually makes other groups seem amateurish. Leaders of the insurance industry attempt to build for the future—in their own interest, of course. As one insurance lobbyist said to me, "Governors and legislatures come and go, but we are always here."

Workers' compensation insurers have a vital stake in continuing the system, and at the same time they know that some changes are necessary. The insurance industry was for a while solidly behind the National Commission's recommended reforms because its leaders were afraid that workers' compensation was becoming so unresponsive to the needs of injured workers that it might become irrelevant. If it did no one any good, it could one day be eliminated by the stroke of a governor's pen.

Insurance company executives are sensitive about the cost issue. They are in the business of selling a service, and like other sellers, they must try to please their customers. The businesses that must buy the insurance are their customers. If they feel they are paying too much for the product they will either agitate for changes at the political level or become self-insured. The larger employers are more and more turning to self-insurance. The loss of the larger companies to self-insurance hurts the compensation carriers, since these are the most lucrative customers, the cream of the crop. The recent election in Ohio with its alliance of business, both large and small, and organized labor against the private insurers shows that business people prefer government-run insurance if it gives them cheaper premiums.

Insurers are caught in a bind. While it is hard to pity the behemoth, he can pity himself, and insurers do. They are caught between the demands of labor for higher benefits and the demands of business for stable premiums. In attempting to meet these two demands without reducing their own huge profits, the insurance industry has taken the lead in many states in trying to change workers' compensation. Insurers like to call their proposals for change "reform." The ultimate goal of these changes however is to keep the system intact and to protect the expense/benefit ratio. The hardest demand for the insurers to square with the demands of business and labor is their own desire to keep the workers' compensation system producing the rivers of money that they themselves are used to swallowing.

Organized labor is no match for the insurance industry in most states. Workers' compensation is only one concern among many facing the labor movement, and not the most pressing one. The needs of its members in a time of economic adversity are many, and labor is hard put to meet them all. In addition, the labor movement on the local and state level suffers from hardening of the bureaucratic arteries. Unable to choose creative new approaches to workers' problems, it waits to defend workers' interests on the battlefield chosen by labor's opponents.

In many states, lawyers' organizations have more political muscle than organized labor. Workers' compensation has become a system of litigation whose volume rivals the criminal justice system. The livelihood of thousands of lawyers depends on its continuence—and its contentiousness. Lawyers know how to organize and they know how to lobby. Furthermore, they understand the importance of spending money for these activities: and they have the money to spend. Although their organizations are much smaller than unions in membership and not nearly as rich as insurance company trade organizations, they nevertheless have a great deal of clout in state capitols. Furthermore, the interests of lawyers representing insurance companies are the same as those representing injured workers—both groups depend on large volumes of litigation. Thus, although the lawyers who defend insurance companies in workers' compensation proceedings cannot openly support their opponents who represent the workers, they tacitly cheer them on, hoping that litigation will not be shut off.

Labor and the insurance industry both understand that litigation should be minimized in workers' compensation. It is, after all, supposed to be a no-fault system. But the insurance industry hurts less from litigation than do injured workers. Therefore, the industry attempts to use its agreement to reduction of litigation as a bargaining chip to get other changes that it wants. Florida in 1979 did what other states could not do. It aced the state's lawyers out of the workers' compensation system. The Florida AFL-CIO, relatively weak in this "right to work" state, and the state's business community agreed on a bill that virtually eliminated litigation from a system that had been rife with it. Part of the package was a cap on premiums, but more significant was the solution to the problem of how to compensate for permanent partial-disability. They simply abolished it! In Florida if you lose your leg you get almost nothing for it, no matter what your employer's negligence, unless you lose some wages, in which case you get some, not all, of the wages lost due to the industrial injury. Also, no matter what happens to you on the job in Florida now, the benefits end in ten years. You can go away and die for all the Florida employers and workers' compensation carriers care.

As bad as the litigation-ridden system of most other states are for the injured workers, the "reformed" system of Florida is surely worse. And yet this type of "reform" is being pushed in many states across the country by a slick insurance-industry public-relations campaign.

It takes some faith to argue that the federal government is the only effective guardian of workers' interests at a moment in history when the reins of that government are held by men whose enmity to labor has been proven. Nevertheless, if there is any hope at all that the instruments of government can be brought to bear to help solve the problems of industrial injury, occupational disease, and workers' compensation, it will be at the federal level, where labor has some influence, not at the state level, where the entire history of the United States has shown that business interests almost always prevail.

10

The Sinister Equation

Industrial injuries and occupational diseases disable and kill—they also cost a lot of money. Who should pay this cost? There are three possible sources: the injured workers themselves, who already must pay the physical cost; the general public through welfare, Social Security, and so on; or the businesses responsible for injuring workers. Workers' compensation was instituted to settle this choice seventy years ago.

Workers' compensation was ostensibly designed as a response to the problem of industrial injury. It was a way of allocating costs. Workers' compensation, it was said, would internalize the costs of injury in the business that caused it. The price of the goods produced by industry would include the cost of its industrial accidents. Therefore, industries that injured many workers would have to charge more for their goods and would be less competitive. The net effect would be a powerful economic incentive for the employer to clean up the workplace.

But workers' compensation does not encourage employers to invest in health and safety. The hidden agenda of workers' compensation, the one area in which it has performed brilliantly, is in insulating employers from liability for lawsuits of their own employees. The result of this is to exempt employers from the strongest economic penalty for putting workers at risk and hasten the socialization of the enormous expense of industrial accident and occupational disease. Instead of internalizing these costs, workers' compensation, by immunizing employers, spreads them around or returns them to injured workers.

In order to truly internalize the costs of workplace injury and disease, workers' rights to sue their employers for negligence must be restored. At the same time, workers' compensation should be truly reformed so that it becomes what it was supposed to be, a no-fault insurance system.

The idea that workers should be allowed to sue their employers sounds like communism to some of the executives of America's corporations, but there is nothing radical about it. Interstate-railroad workers can sue their employers. Their unions have always been strong enough to stave off workers' compensation, despite many attempts by the railroads to foist it on them. The right to bring suit in civil court is one of the oldest and most traditional of all our civil rights. Enshrined in British common law and transplanted to America, it antedates the Bill of Rights by hundreds of years. Cutting off this right between workers and their employers was itself a radical act, and one that has outlived its usefulness.

Those who doubt the efficacy of lawsuits to encourage employers to change their way of doing business should look at the experience of product-liability lawsuits. Since the legal defenses that prevented these suits from being successful were thrown out by the courts and legislatures, manufacturers who never would have considered whether a product they made presented a hazard to the public have become extremely safety-conscious.

Any reform that promotes litigation should be viewed with some suspicion. But allowing workers to sue their employers would actually *reduce* litigation. As the example of product-liability law shows, it is not necessarily large numbers of lawsuits that influence business, but rather the few large lawsuits that demonstrate a real financial threat to a company of ignoring the health and safety of consumers.

Successful managers of the nation's corporations operate under a system of values based entirely on financial considerations. As a general principle, what benefits the company financially is right and what costs the company money with no possibility of return is wrong. Corporate officers' duties are to the balance sheet of the company and nothing else. Those who do not respect this ethic do not rise to top management. Since to waste workers' lives and health cost the company nothing, but changing the conditions that hurt workers does cost money, it would be reckless of corporate managers to dedicate resources to improving conditions. Only the threat of government regulation through OSHA could have any impact, and its effect would depend entirely on the zeal of the government regulators, a zeal notably lacking in the Reagan administration.

The sinister equation must be changed. Wasting workers' lives and health should not equal profits, but losses. If it cost corporations money to injure their workers—enough money to actually influence profits and losses—it would be fiscally irresponsible of management to allow workers to be injured or subjected to potentially disease-producing health or safety hazards.

Let the chemical company which has knowingly exposed its workers to carcinogens be faced with a jury award of $1 million in damages to a bedridden worker, racked with agonizing pain—let this judgment stand and see what the company does to change the conditions of the workplace so that its other workers are not exposed to such hazards. Instead of fighting OSHA, the company will beg OSHA to help find the health hazards and remedy them.

Restoring workers' rights to sue their employers should be balanced by a reduction or elimination of workers' compensation litigation. The system must be reformed so that it can provide at the lowest possible cost what lawsuits cannot—financial first aid for injured workers who cannot prove an employer's negligence.

Most workers' compensation litigation revolves around permanent partial-disability (PPD). When Mike Sarkis injured his back from years of heavy lifting at work, the focus of his case was on the degree of permanent partial-disability that he had suffered.

Under almost all state laws, PPD is evaluated on the basis of physical disability only. Little or no consideration is given to the age, education, or work experience of an injured worker. Martha Cook's disability as a result of her carpal tunnel syndrome was not a major physical disability. After the pain of the operation was over it would not have interrupted the work of a lawyer for an hour. But because her old job as a supermarket clerk could not be done with her injured hand, Martha lost the position she'd had for twenty years. Her working life was devastated. But workers' compensation does not take this into account—she gets almost the same compensation as the lawyer who loses none of his ability to work.

Insurance people love to cite examples of people getting permanent partial-disability awards and then returning to their old jobs, doing exactly the same work they had done before the injury; the workers' compensation laws permit it. But there are far more like Martha Cook who are thrown out of gainful employment with nothing but a token PPD award.

Permanent partial-disability, like the whole workers' compensation system, is mass production of compensation. It was not intended to be tailor-made like the tort system, with its attention to the particular losses of a particular person. Permanent partial-disability in most states is done by numbers. There is a schedule of impairments, and money values are attached to specific impairments. Assessment of impairment of the body as a whole follows guidelines interpreted by the "judge." This approach, legislators once thought, results in a kind of rough justice, an approximation of the economic loss suffered by the injured worker. In fact, there is no justice of any kind in permanent partial-disability. The amounts set in the schedules are totally arbitrary; they reflect not the losses of the injured worker, but the political power of the insurance and business lobby in the state legislature.

Permanent partial-disability on a no-fault basis is what workers got in exchange for giving up their rights to the tort damages of actual wage loss, pain and suffering, and punitive damages. Theoretically, permanent partial-disability compensates only for loss of ability to work. Noneconomic losses are not to be compensated, nor are economic losses that are not the result of inability to work because of an industrial accident or occupational disease. But state legislators have not been able to keep the laws strictly consistent with theory. Indiana's schedule, for example, gives $11,250 to a worker who has lost both testicles on the job. By workers' compensation logic this cannot be justified. The loss of testicles does not by itself mean the loss of ability to work.

While it is illogical to compensate for the loss of testicles, it offends basic notions of justice to deny compensation for such a loss as a matter of principle. The amount of the compensation in this case is like a bad joke, but it is at least a sop to public opinion.

Some states have more than the schedules; they provide wage-loss benefits for victims of permanent partial-disability. However, in many of these states workers are pressured by insurance companies and their own lawyers to settle cases by accepting lump-sum awards that are very attractive in the short run, but do not come close to actually replacing permanent wage loss due to industrial injury. In Massachusetts, for example, lump-sum settlements are the rule although wage-loss compensation is technically available.

Even insurance company spokesmen readily agree that permanent partial-disability is not fair; they acknowledge that workers with serious disabilities are undercompensated while minor or frivolous claims get too much. Florida solved this problem with the Gordian knot approach. That state simply abolished PPD as a category of benefits. There are no schedules, no loss of function awards. Workers who have suffered a permanent disability may collect a wage-loss differential, but that's all. However, even Florida, a "right to work" state dominated by industry, was unable to totally do away with *all* permanent-disability benefits. Florida's legal fig leaf consists of cash benefits for "permanent impairment due to amputation, loss of 80% or more of vision, after correction, or serious facial or head disfigurement." A worker who suffers any of these injuries on the job is entitled to $50 for each degree of disability up to 50% and $100 for each degree of disability above that. With its new "reformed" law, a Florida worker whose legs are cut off when a forklift falls on him is entitled to $7500 in cash.

Some insurance spokesmen have called permanent partial-disability "blood money," and they have a point. Even Florida's extreme antiworker workers' compensation law includes a token payment for severe injuries. There is still a bare reminder in this kind of statute that there once was a kind of bargain; that workers had a very important right that was given up in order to get something else. It is not yet time, industry knows, to eliminate the last remnant of what workers bargained for—permanent partial-disability on a no-fault basis. But the Florida law shows the way industry and the workers' compensation insurers are thinking about changing the system in the future. As bad as workers' compensation is now for workers, it will get much worse if other states follow Florida's example.

Permanent partial-disability is workers' compensation's most litigated benefit. It doesn't take care of the workers who need it the most and it sometimes overcompensates those that need it the least. But to abolish it, as Florida effectively did, is to renege on the "historic bargain" by which workers' rights to sue were given up. That bargain must not be forgotten. It was bad enough for workers when the only quid pro quo for permanent impairment was PPD. Without even that, workers' compensation becomes a clear-cut case of oppression.

If a worker's right to sue his employer were restored, the need for permanent partial-disability would be gone. Its abolition would result in a tremendous reduction in workers' compensation litigation that would more than offset the increase in civil lawsuits against employers.

A truly reformed workers' compensation system should provide a disabled worker with wage-loss benefits on a no-fault basis. Compensation for physical impairment and for pain and suffering should be left to lawsuits based on negligence.

Occupational Disease—Abolish the Threshold

Whatever is done or left undone with permanent partial-disability will have little effect on victims of occupational disease. At least 95% of all occupational-disease cases never cross the threshold of the workers' compensation system; they are never accepted as work-related.

Work-relatedness is the critical issue with occupational disease because scientific evidence of causation of most diseases is just not advanced enough to satisfy legal challenges within the workers' compensation litigation system.

In this system business and the insurance industry always have the upper hand, particularly in complex litigation involving extremely technical medical issues. Workers' compensation "judges" tend to be conservative—they don't like to break new ground. The ability of science to conclusively establish the work-relatedness, particularly of nontraditional occupational diseases like cancer, is many years away. Most evidence is based on animal tests and is thus subject to dispute. With traditional occupational diseases such as byssinosis, the problem is the shortage of doctors able to properly diagnose individual cases.

Some state officials and members of Congress are attempting to find a solution to the problem of how to compensate occupational disease. One approach that has been suggested deals with asbestos victims. The idea is to institute a presumption of compensability similar to the presumption that allows miners compensation for black lung disease. This proposal has enthusiastic backing of the asbestos-goods manufacturers. Why? In exchange for their backing they have been promised that the federal government will shoulder a large part of the financial burden and, even more important, the right of "third parties" to sue for asbestos disease they contracted due to the manufacturers' negligence would be cut off. The manufacturers will oppose any reform proposal that does not clear them of all responsibility for their actions. They hope to hold hostage the hundreds of thousands of asbestos victims, saying, in effect, to organized labor and its liberal sympathizers: "These asbestos victims will starve if you don't give in to our demands."

In an era of limits it seems wrong that public money should be dedicated to paying what is really industry's bill. More important is the danger to workers of further insulating manufacturers from the consequences of their lack of concern for their workers' health and safety. Transcending the cold question of costs is the danger of suffering and death of millions of American workers of current and future generations. As with industrial injury, employers must be financially penalized for knowingly or negligently endangering their workers' health.

Other reforms are being studied. But they all run into the problem of work-relatedness—how can it be established? The solution is to abolish the work-relatedness test altogether. It is a relic of an earlier time.

Lawrence Baker, president of Argonaut Insurance and a maverick among insurance executives, said to me, "I see no need for maintaining a differentiation for work-related injury. The real issues are what level of income maintenance should be provided and what kind of health care." The work-relatedness issue distracts from these really essential issues, Baker believes. John F. Burton, former chairman of the National Commission on State Workmen's Compensation Laws, has also suggested that the work-relatedness test be dropped, at least for occupational disease, among which he places most back strains and sprains. *

Compensation for disease or disability based solely on the severity of the condition, without regard to its cause, is, of course, the general rule of health and disability insurance—except workers' compensation. Seventy-five percent of American workers are already covered by group health insurance. † For these, the medical portion of workers' compensation is not needed. In fact, this aspect represents double coverage, paid by the nation's employers and workers for the same coverage.

Income-maintenance insurance is less widespread, but still fairly common. Usually called long-term disability (LTD), this type of insurance is not concerned with the causes of disability, but simply with whether the insured is able to work or not. Forty percent of all American workers are covered by this type of insurance at no cost to themselves. ‡

Clearly, some type of no-fault insurance scheme is needed to provide medical care and income maintenance for workers injured on the job. And this insurance, like the present workers' compensation, should be mandatory and paid by employers. As the National Commission on State Workmen's Compensation Laws recommended, the coverage should be very broad—it should cover every American worker. It is the lowest-paid and least-educated workers who need income maintenance and health care the most.

The one benefit unique to workers' compensation, and the stum-

*John F. Burton, "The Challenge of Diseases for Workers' Compensation." Speech presented at the Seminar on Current Issues in Occupational Disease, sponsored by the National Council on Compensation Insurance, New York, Oct. 7, 1981 (unpublished).

†*The National Underwriter*, Life and Health Insurance Ed., July 4, 1981, p. 1.

‡Id.

bling block to reform, is permanent partial-disability. If that were abolished and the workers' right to sue restored, workers' compensation could become what it should be—an insurance system instead of a litigation system.

Arguing, as many insurers do, that more time is needed to study the problem is indecent. Workers' compensation has been around for seventy years and it has never internalized the costs of injury and disease. The millions of workers at risk can no longer wait.

While it is critical for workers to have the right to sue, a "private right of action," government regulation of the workplace is also essential. A renewed OSHA, dedicated to protecting American workers, should be a central objective of all those who believe in a just society. There should be no doubt that the present administration will try to slowly suffocate OSHA. It must not be allowed to succeed.

Underlying all proposals for concrete reforms and institutional changes is a need for strong political support for workers' health and safety. Laws and agencies can never by themselves better the lives of our citizens. Only an aroused and educated public can ensure that progress in technology and methods of production will be matched by a corresponding progress in the conditions of work and compensation for injury and disease.

Bibliography

Agran, Larry. *The Cancer Connection*. Boston: Houghton Mifflin, 1977.

Ashford, Nicholas. *Crisis in the Workplace: Occupational Disease and Injury*. Cambridge, MA.: MIT Press, 1976.

Barth, Peter S., with Hunt, H. Allan. *Workers' Compensation and Work Related Illnesses and Diseases*. Cambridge, MA.: MIT Press, 1980.

Berkowitz, Monroe. *Workmen's Compensation: the New Jersey Experience*. New Brunswick, New Jersey: Rutgers University Press, 1960.

Berman, Daniel M. *Death on the Job: Occupational Health and Safety Struggles in the United States*. New York: Monthly Review Press, 1978.

Boyd, James Harrington. *Workmen's Compensation and Industrial Insurance Under Modern Conditions*. Indianapolis: Bobbs-Merrill, 1913.

Brodeur, Paul. *Expendable Americans*. New York: Viking Press, 1974.

Brown, Michael. *Laying Waste, The Poisoning of America by Toxic Chemicals*. New York: Pantheon Books, 1979.

Cheit, Earl. *Injury and Recovery in the Course of Employment.* New York: Wiley, 1961.

Cheit, Earl, and Gordon, Margaret S., Editors. *Occupational Disability and Public Policy.* New York: Wiley, 1963.

Chelius, James Robert. *Workplace Safety and Health: The Role of Workers' Compensation.* Washington: American Enterprise Institute for Public Policy and Research, 1977.

Clapham, J. H. *Economic History of Modern Britain.* Cambridge, England: The University Press, 1926–1938.

Commons, John R. *History of Labour in the United States, 1896–1932.* New York: Macmillan, 1918–1935.

Dawson, William Harbutt. *Social Insurance in Germany.* London: T. Fisher Unwin, 1912.

Dodd, Walter F. *Administration of Workmen's Compensation.* London: Oxford University Press, 1936.

Eastman, Crystal. *Work Accidents and the Law.* New York: Charities Publication Committee, 1910.

Evans, Roy R. *Tragedy at Work.* Austin, Texas: Futura Press, 1979.

Follman, Joseph F. *The Economics of Industrial Health.* New York: AMACOM, 1978.

Garvin, James Louis. *The Life of Joseph Chamberlain.* London: Macmillan and Co., 1935.

Gully, Elsie E. *Joseph Chamberlain and English Social Politics.* New York: Columbia University Press, 1926.

Hanes, David G. *The First British Workmen's Compensation Act.* New Haven: Yale University Press, 1968.

Hanna, Warren. *California Law of Employee Injuries and Workmen's Compensation.* Albany, New York: M. Bender and Co.

Kirby, John. "Cruel Unionism." National Association of Manufacturers Pamphlet #20. Speech delivered at Kenyon College, Gambier, Ohio, on February 3, 1911.

Kochan, Thomas. *Effectiveness of Union–Management Safety Committees.* Kalamazoo, Michigan: W. E. Upjohn Institute for Employment Research, 1977.

Larson, Arthur. *The Law of Workmen's Compensation.* Albany, New York: M. Bender and Co., 1952.

Lehmann, Phyllis. *Cancer and the Worker.* New York: New York Academy of Sciences, 1977.

Link, Arthur S. *American Epoch, 3rd Edition*. New York: Knopf, 1967.

Page, Joseph A., and O'Brien, Mary Win. *Bitter Wages*. New York: Grossman Publishers, 1973.

Roberts, Maurice G. *Injuries to Interstate Employees on Railroads*. Chicago: Callaghan and Co., 1915.

Schwedtmann, Ferd. "Co-operation or What?" National Association of Manufacturers Pamphlet #38. Speech delivered to the Second Annual Convention of the International Association of Casualty and Surety Underwriters, August 14, 1912.

Schwedtmann, Ferd, and Emery, James A. *Accident Prevention and Relief. An Investigation of the Subject in Europe with Special Attention to England and Germany*. New York: National Association of Manufacturers, 1911.

Somers, Herman Miles, and Somers, Anne Ramsey. *Workmen's Compensation: Prevention, Insurance, and Rehabilitation of Occupational Disability*. New York: Wiley, 1954.

Spender, J. A. and Asquith, Cyril. *Life of Herbert Henry Asquith, Lord Oxford and Asquith*. London: Hutchinson and Co., 1932.

Stellman, Jeanne. *Women's Work, Women's Health: Myth and Realities*. New York: Pantheon Books, 1977.

Villard, Harold Garrison. *Workmen's Accident Insurance in Germany*. New York: Workmen's Compensation Publicity Bureau, 1913.

Whiteside, Thomas. *The Pendulum and the Toxic Cloud: The Course of Dioxin Contamination*. New Haven: Yale University Press, 1979.

Wilson, Sir Arnold, and Levy, Prof. Herman. *Workmen's Compensation*. London: Oxford University Press, 1939 (V.1.), 1941 (V.2).

Yellowitz, Irwin. *Labor and the Progressive Movement in New York State, 1897–1916*. Ithaca: Cornell University Press, 1965.

Selected United States Government Publications

"Compendium on Workmen's Compensation." National Commission on State Workmen's Compensation Laws. Washington: U.S. Government Printing Office, 1973.
"An Interim Report to Congress on Occupational Diseases." U.S. Department of Labor, Assistant Secretary for Policy Evaluation and Research. Washington: U.S. Government Printing Office, 1980.
"Lost in the Workplace: Is There an Occupational Disease Epidemic?" Proceedings from a seminar for the news media, September 13–14, 1979. Washington: U.S. Government Printing Office, 1979.
"Report." National Commission on State Workmen's Compensation Laws. Washington: U.S. Government Printing Office, 1972.
"Supplemental Studies for the National Commission on State Workmen's Compensation Laws." National Commission on State Workmen's Compensation Laws. Washington: U.S. Government Printing Office, 1973.

Index